ESTIMATING FOR BUILDERS AND SURVEYORS

ESTIMATING FOR BUILDERS AND SURVEYORS

Hugh Enterkin M.Sc., F.R.I.C.S., F.I.O.B., F.I.Arb.
Senior Lecturer
Department of Surveying and Building
Dundee College of Technology

Gerald Reynolds M.Sc., F.R.I.C.S., M.I.O.B.
Principal Lecturer
Department of Surveying and Construction
Newcastle-upon-Tyne Polytechnic

HEINEMANN : LONDON

William Heinemann Ltd
15 Queen Street, Mayfair, London W1X 8BE

LONDON MELBOURNE TORONTO
JOHANNESBURG AUCKLAND

First published 1972
Second edition 1978
434 90542 9

Made in Great Britain at The Pitman Press, Bath

Preface

This book has been written to meet the requirements of students pursuing courses leading to the examinations of:

1. Royal Institution of Chartered Surveyors
2. Institute of Quantity Surveyors
3. Institute of Building
4. Ordinary and Higher National Diplomas
5. Higher National Certificate
6. City and Guilds of London Institute

We trust it will also be of interest to small and medium-sized builders, to estimators and to practising surveyors generally. The aim of the book is to provide factual data by means of the detailed analysis of prices for those items commonly used in current building practice and to illustrate this by providing the optimum of practical examples. It also shows methods of dealing with preliminaries and *pro rata* rates; and it illustrates the effect of cash flow in respect of building contracts.

Briefly, the changes are: the Introduction has been amended, and the chapters on the 'All-in' Rate is new. This chapter has been re-written to meet the current (1977) regulations regarding payments to artisans and labourers in Building Industry and the wage rates are based on the wages agreement which became operative in June 1977 for building trade operations, except plumbers. The plumbers' rates are based on the agreement which became operative during August 1977. The new edition also meets with the requirements of the Sixth Edition of the S.M.M. generally.

The chapters on Mechanical Plant and Preliminaries have been extensively adjusted to comply with current (1977) practice whilst Chapters 5 to 18 inclusive have been amended and updated in accordance with materials costs as at April/May 1977 and wage rates as described in the foregoing paragraph referring to 'All-in' Rates; and examples have been added where appropriate. Chapter 19, Drainage, has been completely re-written conforming with modern techniques applying to drainage whilst the chapters on Pro-Rata rates and Examination questions are new.

Our readers will appreciate that all matters of legislation, working conditions, wages, and the like are subject to change, and the latest official information should be consulted in actual practice.

We wish to express our thanks to M. Bone for his assistance.

We also wish to express our thanks to The Royal Institution of Chartered Surveyors, The Institute of Quantity Surveyors, and The Scottish Technical Education Council (SCOTEC) for allowing the examination questions to be reproduced at Chapters 20 and 21.

Dundee H.E.
G.R.

Contents

1 Introduction

ESTIMATING

Estimating is a system of 'building-up' or compiling notes to facilitate competitive tendering and is the technical process of predicting costs of construction.

Any person who wishes to spend money on building has many ways open to him to obtain tenders. The following indicates the main approaches:

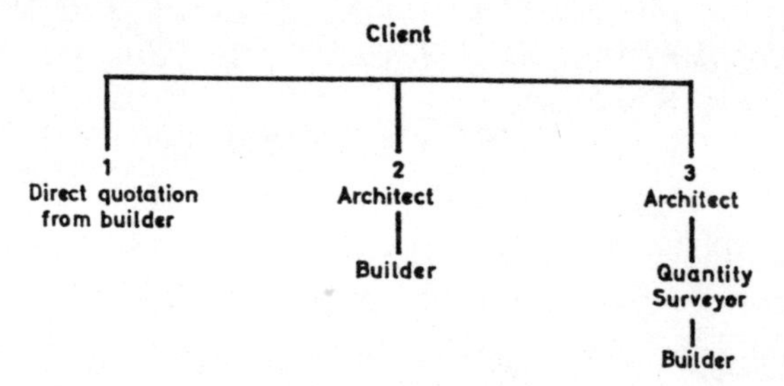

1. There are normally no contract documents, except a letter from the client to the builder inviting him to provide an estimate for the work as described in the letter or from an inspection on the site.

   The most common form this takes is, say, a painter enters a room, is told by the client how he wants the walls, doors, etc. treated, and offers to carry out the work for a certain amount.

2. This method is where the client appoints an architect who prepares a set of plans and a specification of the type of materials and the scope of the work to be tendered for. The plans and specifications are then sent direct to the builder, who prepares a quotation for the work involved.

   This method is commonly used despite having the obvious disadvantage of there being no common document for two or more builders from which to price.

3. This is the most usual method for contracts of any size

from £1,000 and upwards. Contracts expected to be over £8,000 are normally prepared with a Bill of Quantities. Under this amount it is felt that a contract without quantities but including a specification with a description of works is sufficient. (This of course could be done under 2.) The Specification, etc., may be prepared by the Quantity Surveyor or the Architect.

The client appoints an architect who prepares the plans. The quantity surveyor (if we can ignore the function of cost planning) then prepares a Bill of Quantities which is sent to a number of selected contractors for pricing.

The rates in this book are prepared on the assumption that Method 3 is used and that the contract is based on the current version of the *Standard Conditions of Contract*, 1963 edition, and the *Standard Method of Measurement of Building Works*, Sixth Edition (S.M.M.).

Section A (General Rules) of the S.M.M., paragraph A4.2, states:

'Unless otherwise specifically stated in the bill, or herein, the following shall be deemed to be included with all items:

(i) Labour and all costs in connection therewith.
(ii) Materials, goods and all costs in connection therewith (e.g. conveyance; delivery; unloading; storing; returning packings; handling; hoisting; lowering).
(iii) Fitting and fixing materials and goods in position.
(iv) Plant and all costs in connection therewith.
(v) Waste of materials.
(vi) Square cutting.
(vii) Establishment charges, overhead charges and profit.'

We have therefore allowed for profit and oncosts in each item in this book to comply with A4.2.(g).

DECISION TO TENDER

Once a builder receives either an invitation to tender (which is good practice) or the tender documents themselves, the first question is: Does he wish to tender for the job?

The answer to this will depend on a number of factors:

1. Does he wish to carry out work for the client (or his representatives)?
2. Can the firm's resources match the work load, bearing in mind the amount of work in hand? The contract may, for example, create an imbalance of work between trades which may be overcome by sub-contracting (if allowed).
3. Is the job in an area posing special problems? It might, for example, mean setting up a camp for employees within reasonable travelling distance.
4. What type of competition does he face? If he knows that a fellow contractor needs a job for turnover and is normally quite competitive, he may decide that it is not worth while having his estimator prepare a tender that is not likely to be the lowest.
5. How many jobs are in the pipeline? He may decide to allow his competitors to obtain this particular contract in the hope that the competition will not be so fierce for the next contract.

The answers to these questions can depend as much on geographical position as on the number of contracts being entered into. That is, in an area with only a limited number of contractors interested in a type of contract which is on the increase (e.g. alteration work to existing houses receiving grants), the builder can virtually pick and choose at his own price.

Having decided to tender, he should have the documents inspected by the estimator to ensure that all the required information is given.

1. What is the type of contract?
2. What are the restrictions? These include time, position and access to site, order of the works, sectional completion, liquidated and ascertained damages, retention, form of payment, etc.
3. What amount of work can be handled by his own firm and what amount will have to be sub-contracted?
4. What is the amount of Prime Cost and Provisional Sums (indicating the number of nominated sub-contractors and suppliers and how well the documents have been pre-planned)?
5. Can the item descriptions be priced, i.e. do they

describe adequately the work involved?

6. Are any of the various contract documents contradictory?
7. Are there any alterations or deletions of standard clauses or supplementary clauses?

VISIT TO SITE

The estimator will require first-hand information and if he does not visit the site himself will send, depending on the size of the job, his foreman, site agent, contracts manager, etc., who will take notes, normally on a *pro forma* (*see* Appendix I to this chapter).

QUOTATIONS

Meanwhile, the estimator will have 'broken up' the bill into the sections which he will price and those for which it will be necessary to obtain quotations from sub-contractors (*not* nominated sub-contractors) normally chosen by him but for which he must receive the architect's permission before placing a firm order. It is often the practice for a page of the Bill of Quantities to be set out for the contractor to enter in the names of the firms carrying out the various sections of work.

The estimator will abstract the required information from the Bill of Quantities and pass it on to the sub-contractors (*see* Appendix II this chapter). The lowest acceptable quotation is then priced into the Bill of Quantities with sometimes a percentage addition, either in the rates or at the end, to cover profit and attendance.

NOMINATED SUB-CONTRACTORS

The architect has the right under the Standard Conditions of Contract to nominate such firms as he deems to be necessary for the proper execution of the contract. These firms are known as nominated sub-contractors and their position in the contract is clearly defined under Clause 27. They can be furniture contractors in one contract and steelwork contractors in another. A Prime Cost or Provisional Sum is included in the Bill of Quantities or Specification and the main contractor is instructed, unless he has valid reasons otherwise, to accept the nominated firms as nominated sub-contractors. Within the contract the main contractor is entitled to $2\frac{1}{2}\%$ cash discount, which is normal trade practice to encourage early payment, and to a profit item and a general attendance item (defined in S.M.M. B9.2).

The profit item is normally a percentage addition and if not will be *pro rata* to the Prime Cost Sum if different in the final account.

The general attendance item is a lump sum which is not altered unless the nominated sub-contractor's original contract is considerably changed (e.g. half the number of windows).

Special attendance on nominated sub-contractors must be given as detailed items and the estimator will price accordingly (B9.3).

NOMINATED SUPPLIERS

An architect may nominate suppliers for a variety of reasons:

(*a*) The client may not be able to make up his mind, e.g. on ironmongery.

(*b*) The trade may change its styles every year, e.g. of wallpaper, electric light fittings.

(*c*) He is not as far on with planning the job as he should be. Again a Prime Cost or Provisional Sum is included in the Bill of Quantities or Specification and an item for profit allowed. The main contractor is allowed 5% discount but, unlike nominated sub-contractors, cannot hold a retention. The term 'nominated supplier' is also clearly defined under Clause 28.

Fixing the goods and materials is measured and items are included in the section of the bill where required. The cost of unloading, storing, hoisting the goods and materials and returning packing cases, etc., carriage paid, and obtaining credits for them is included in the item for fixing.

PROJECT APPRECIATION

It is helpful for the estimator to meet the people concerned with the design aspects of the job and he should visit the architect and any other of the consultants he considers necessary at an early date. He should inspect the drawings, site reports and any other available information and assess the state of the design. A design which is not properly planned will be the cause of delays and disputes and an allowance will have to be made for this in the tender, even although there are certain safeguards within the Standard Conditions of Contract.

When the estimator has become acquainted with the project, the manner in which the job will be carried out should be discussed with those who will be directly responsible, such as the site agent, for the programme of construction and for plant, etc. if the tender is accepted. A programme of work sequences should be made up, and labour site staff, plant and facilities decided before calculating the individual rates.

If the tender is accepted or if the client requires to see the programme of work before accepting the tender, these discussions will be the basis on which the programme will be prepared.

SUBMISSION OF TENDER

The main contractor should re-assess the conditions of contract, quotations, work load and prevailing market conditions, etc. before final submission of tender. The prices put into a Bill of Quantities are quite often copied by an office junior, and, especially where the tender is subject to adjustment after checking the bill, this should be checked by the estimator.

A record should be kept of all tenders submitted and an analysis made, whether successful or not, when the list of tenderers submitted, together with tender amounts, is made known. These lists are normally made up in alphabetical order of the names of the firms, with a subsequent statement on the tender amounts listed in order of lowest to highest (Scottish practice). Thus the name of the firm appearing first on the list should not necessarily be bracketed with the lowest tender.

PROFIT AND ONCOSTS (OR OVERHEADS)

Profit is the amount left over after all outgoings have been charged and is usually related in percentage to the amount of capital invested in the business. If you compare this percentage with the return given by building societies, local authorities and other secure forms of investment, it will become apparent whether the effort is worth while. Most builders today are looking for approximately 10 to 12½% return on their capital investment.

The amount of profit will depend amongst other considerations on the size of the business, turnover, market conditions, and nature and type of work being tendered for.

Oncosts or Overheads

These can be listed as follows:

(1) Head Office charges.
(2) Site staff etc.
(3) Overtime not specifically ordered by the architect.
(4) Time lost due to inclement weather.
(5) Bonuses and other incentive payments.
(6) Apprentices' study time etc. (Construction Industry Training Board).
(7) Employer's contribution to National Insurances, including Graduated Pensions.
(8) Selective Employment Tax, etc.
(9) Annual and public holiday costs.
(10) Fares, travelling time, etc.
(11) Subsistence allowances.
(12) Safety and Welfare facilities.
(13) Third party and employers' liability insurances.
(14) Sick pay.
(15) Redundancy Payments Act 1965 and 1969, Employment Protection Acts.
(16) Tool allowances.
(17) Use etc. of small tools.
(18) Staging, trestles, artificial lighting, etc.

The above list is not exhaustive and the builder would include in the list any other liabilities or obligations he considered necessary.

The overheads are priced under a number of headings in a Bill of Quantities, e.g. items 2, 12, 13 and 18 would probably be charged in the Preliminaries section whilst items 3 to 11 and 14 to 16 would be included in the 'all-in' rate, although it is the practice of some firms to price for a few of these in the Preliminaries section and not in the rate.

If the work is carried out under daywork in a building contract, the overheads may be defined to include for items 1 to 18 with a percentage being added to standard time, etc. rates to arrive at the amount to be charged.

Thus, each firm may have its own interpretation of overheads and the percentage figure used by firm A may not relate to the percentage used by firm B.

Oncosts can only be calculated by accurate costing and record-keeping over a considerable period of time, and for the purposes of this book will include for the following:

(1) Light plant.
(2) Salaries of office staff.
(3) Buildings: rents, rates, maintenance, etc.
(4) Office expenses: stationery, stamps, etc.
(5) Telephone accounts, fire insurance and other periodical expenses.
(6) Fuel and light.
(7) Motor cars, if used for business purposes.
(8) Money costs: interest on overdrafts, etc.

These oncosts are normally stated as a percentage based on turnover, e.g. turnover for the year 1977 was £2,000,000 with oncosts coming to £150,000; this would give a percentage of:

$$\frac{50{,}000}{2{,}000{,}000} \times 100 = 7\tfrac{1}{2}\%$$

For the purpose of all examples to be worked in this book a combined percentage of 20% will be used for profit and oncosts, although it must be recognised that this figure will vary according to the demand for work.

Thus, to build up individual rates the following must be taken into consideration:

(1) Materials:	S.M.M. A4.2(b)(d)(e) and (f);
(2) Labour:	S.M.M. A4.2(a) and (c);
(3) Establishment and Overhead Charges:	S.M.M. A4.2(g); and
(4) Profit:	S.M.M. A4.2(g).

PROFIT AND CASH FLOW

What is profit? Profit is the surplus of income over expenditure. However, the profit shown in the accounts of a firm is not necessarily reflected by an equivalent increase in cash. One of the reasons for this is that, in arriving at the profit for the year, a firm includes in its expenses an amount for depreciation in respect of its fixed assets, e.g. buildings, plant, vehicles, etc.

The cash which was used to purchase these assets was spent some time in the past, so the depreciation charge in the Profit and Loss Account does not represent a current outgoing of money. Consequently, the cash coming into the firm is the net profit shown by the Profit and Loss Account plus the amount charged by way of depreciation. The amount of money representing the depreciation charge is usually retained within the firm as additional working capital.

Profit on its own does not mean a great deal. It must be compared with some significant factor within the business, e.g. net capital employed (i.e. total assets less current liabilities) or with the shareholders' interest in the business (i.e. share capital and reserves), to give an indication as to whether the business is earning an adequate return.

In the same way comparison can be made in connection with projects to see whether the project earns sufficient return on the investment. A more modern approach is to make use of Discounted Cash Flow techniques where all future profits are discounted to a present value. The resultant present value is then compared with the outlay or capital invested in the project to give an indication as to the profitability or otherwise of the project; alternatively, the present values of rates of return can be compared in order to rank different projects in order of profitability.

Where a contract runs for a number of years, it is necessary for the builder to obtain cash from the client at regular intervals to maintain an adequate amount of working capital to enable him to carry on with the contract. The terms of the contract will state what percentage of work done is to constitute the progress payments or instalments, and what is to be held by the client as retention money.

From the builder's point of view it is important to receive these payments promptly and regularly; otherwise his cash position will deteriorate and he may have to resort to borrowing, thus incurring additional interest charges.

The use of a cash budget drawn up in respect of each project and for the business as a whole is an advantage. This is best illustrated by means of an example.

A contract was entered into for the erection of a four-storey building having a contract sum of £255,500. The JCT Standard Conditions of Contract set out the various particulars some of which were:

Defects Liability Period	6 months.
Date for Completion	18 months from Date for Possession.
Liquidated and Ascertained Damages	At the rate of £250 per week or part thereof.
Period of Interim Certificate	1 month.
Retention Percentage	5%.
Period of Final Measurement and Valuation	6 months from Date of Practical Completion.

The Final Certificate showed a Contract Sum of £260,000, the contract period having extended to 25 months. The extra £4,500 being the amount ascertained under the various clauses of the conditions of contract as being due to the contractor.

A graph showing the anticipated expenditure build-up was drawn and the amounts, as stated in the Certificates of Payment of work done were marked on for an 18-month contract period, the graph for the 25-month period being drawn at a later date. This graph was intended not only to assist the employer in an assessment of his cash flow but also to give an indication as to whether the contractor was keeping up to his programme of work.

The graph appears to indicate monthly instalments missing. This is brought about by some Certificates of Payment being passed at the end of one month with the following Certificate of Payment being passed at the beginning of the next month but one.

The figures plotted on the graph would seem to indicate that the work as carried out had little relationship to the original programme.

In the schedules relating to the contract graph (*see* Tables A, B and C) one can see that the cash budget of receipts over the original eighteen months was £249,112 (Table B). This figure can be split down month by month to be incorporated into the cash budget for the business as a whole.

The advantage of the cash budget is that it enables a business to see in advance when surpluses or shortages of cash are likely to occur and the business can plan accordingly.

As a result of the contract extending to 25 months the cash position would deteriorate since instead of receiving £249,112 in 18 months with the balance of retention money coming 6 months later (i.e. in the 24th month), the builder received £208,050 in 18 months. The remaining £51,950 would be received in 2 lots, £45,450 over the months extension of the contract and the balance of the retention money £6,500, 6 months after that.

In addition to a possible loss of £4,500 on the contract, unless the contractor can justify additional claims (i.e. value of work done £260,000, Fixed Contract Price £255,500) the contract would incur additional losses due to having to main-

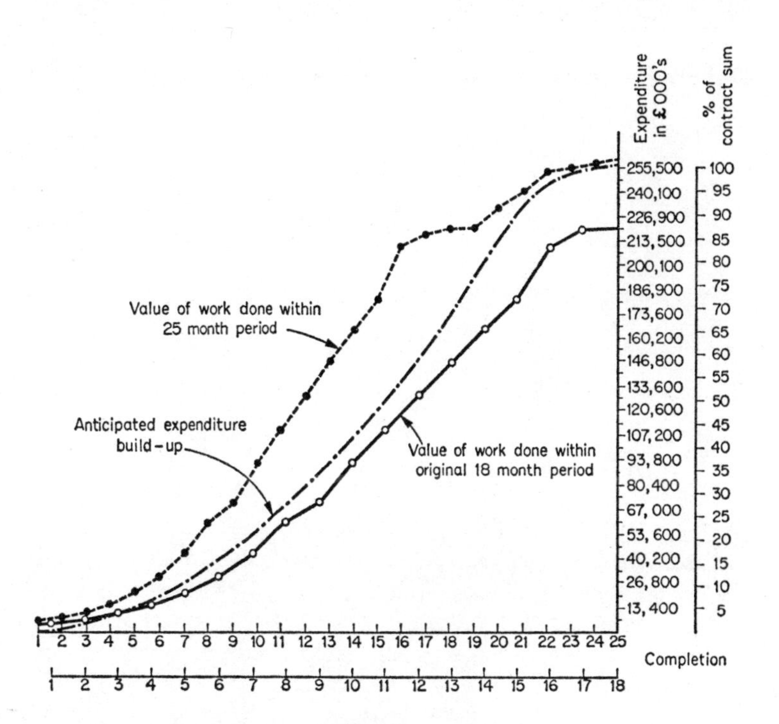

tain a site agent, plant, huts, etc., on the site for 7 months longer, when they could have been employed on other contracts. These are opportunity costs and if the cost of a site agent and plant and so on was £200 per week, this would amount to £6,000 over the 7 months.

Adding to the builder's losses on the contract might be the clause for liquidated and ascertained damages of £250 per week which in turn would amount to £7,500 over 7 months. In fact this figure was not established until final account agreement.

TABLE A

Actual Performance of Contract from Graph

Month	Value of Work Done £	Cumulative Amount of Work Done	5% Retention £	Cash In Flow £	Cumulative Amount of Cash Flow
1	3,350	3,350	168	3,182	3,182
2	3,350	6,700	167	3,183	6,365
3	3,300	10,000	165	3,135	9,500
4	5,200	15,200	260	4,940	14,440
5	6,800	22,000	340	6,460	20,900
6	8,800	30,800	440	8,360	29,260
7	13,900	44,700	695	13,205	42,465
8	15,600	60,300	780	14,820	57,285
9	10,050	70,350	502	9,548	66,833
10	21,750	92,100	1,088	20,662	87,495
11	18,400	110,500	920	17,480	104,975
12	16,800	127,300	840	15,960	120,935
13	17,700	145,000	885	16,815	137,750
14	18,200	163,200	910	17,290	155,040
15	18,800	182,000	940	17,860	172,900
16	27,500	109,500	1,375	26,125	199,025
17	8,000	217,500	400	7,600	206,625
18	1,500	219,000	75	1,425	208,050
19	1,200	220,200	60	1,140	209,190
20	10,800	231,000	540	10,260	219,450
21	8,000	239,000	400	7,600	227,050
22	15,000	254,000	750	14,250	241,300
23	2,000	256,000	100	1,900	243,200
24	3,000	259,000	150	2,850	246,050
25	1,000	260,000	50	950	247,000
	260,000		13,000	247,000	
	Practical Completion		6,500	6,500	
	End of Defects Liability		Nil	6,500	260,000

TABLE B

	Value of Work Done £	Retentions (payable 6 months after completion) £	Cash Flow £
Planned for 18 months	255,500	6,388	249,112
Actual for 18 months	219,000	10,950	208,050
Actual for 25 months	260,000	6,500	260,000

TABLE C

Using Discounted Cash Flow techniques, to show the difference in Net Present Values between the budgeted outgoings on the contract over 18 months, and the actual outgoings over 25 months, taking into account retention money.

Budgeted Situation
(Read from 'anticipated expenditure build-up' curve)

Months	Incomings	Present Value Factor at $12\frac{1}{2}\%$	Net Present Value
12	£145,000	0·8889	£128,890
18	£104,112	0·8395	£ 87,402
24	£ 6,388	0·7901	£ 5,046
	£255,500		£221,338

Actual Situation

Months	Incomings	Present Value Factor at $12\frac{1}{2}\%$	Net Present Value
12	£120,935	0·8889	£107,390
18	£ 87,115	0·8395	£ 73,133
24(25)	£ 45,450	0·7901	£ 35,910
30	£ 6,500	0·7462	£ 4,850
	£260,000		£221,283

These cash flows are based as from the commencement of the contract and indicate that in present value terms there was not a great difference overall due to the contract extending to two years.

The $12\frac{1}{2}\%$ rate is the rate of return required by the builder on the contract.

The figure of £260,000 includes the additional £4,500 included in the Contract at a later stage. The present value of £4,500 at this time is found to be £3,358. Therefore for comparison purposes, the Present Value is £221,283–£3,358, i.e. £217,925.

Also the present value of the additional site costs of £6,000 is found to be £4,477 and the present value of the liquidated damages is £5,926.

Although the Table C shows only a difference of £55 in Present Value terms, to this there has to be allowed:

(*a*) Additional Contract Costs	£4,500 the present value of which is (4,500 x 0·7462)	£3,358
(*b*) Additional Site Costs	£6,000 the present value of which is (6,000 x 0·7462)	£4,477
(*c*) Liquidated Damages	£7,500 the present value of which is (7,500 x 0·7901)	£5,926
		£13,761
Less difference as per Table C		£ 55
	Loss on Contract in Present Value Terms	£13,706

If the builder realized that the contract could not be completed within the 18 months envisaged then, ignoring the liquidated damaged factor, his tender should have been increased to approximately £266,000 (£255,500 + £4,500 + £6,000 as above).

The original profit (i.e. 12½%) on the cost of £221,338 would be £27,667. This is reduced by the additional costs and damages to £13,961 and so has the effect of reducing the profit by approximately 50%.

N.B. The additional Contract Costs only takes into account the amount considered allowable under the conditions of contract. The contractor's additional outgoings, in relation to contract costs, may have been considerably more than the figure of £4,500 paid.

APPENDIX I

Pro Forma Site Visit Report

A. BUILDER & SON (BUILDERS) LTD.
Bell Street, Dundee, Scotland.

Site Visit Report By ....

Completed by Estimating Dept.		
1. Name of contract		
2. Location		
3. General description of contract		
4. Drawings available for inspection at		
5. Contract particulars, viz: (a) Architect	Name	Address with Tel. No.
(b) Engineer		
(c) Quantity Surveyor		
(d) Date of possession of site		
(e) Date for completion		
(f) Ascertained and liquidated damages		
6. Date for return		
7. Names of Local and Statutory Authorities		
8. Availability of Materials		
9. Weather conditions (if high rainfall etc.)		

Site visited by		Date of visit			Distance from area office/yard......mls		
Labour	Proposed working hours per week	Site Staff (if any)					
	Availability of Labour and any additional hourly incentive payments (excluding bonus) to either obtain or retain labour	Trade	Local	Plus Rate	Imported	Plus Rate	Remarks
		Bricklayers Joiners Slaters Roughcasters Plumbers Plasterers Painters					
	Transportation of Labour	Yes/No	Method to be used and Fares (if any) to be paid				
	Lodging allowances	Yes/No	Details of camp accommodation or lodging facilities				
Site Conditions	Nearest Tip from site	....mls	Charges for use ....per load				
	Ground conditions strata expected to be encountered	Rock yes/no	Clay yes/no	Running Sand yes/no	Boulders yes/no	Close sheeting Excavations below what depth if any	...m
	Water	Surface yes/no	Spring yes/no	Tidal yes/no	Pumping yes/no	Dewatering yes/no	State number of weeks
	Site access Temporary roads Existing roads Access difficulties	State requirements if any					
General Requirements	Is site fenced	yes/no	Will watchmen be required		yes/no	full-time/ part-time	
	Will hoarding be required	yes/no	If yes, give	Length Height			
	Site Storage required e.g. compound, special stores, etc.						
	Any other special requirements						
Services	State distance and availability of water, gas, electricity and sewers						
Plant	State of Plant requirements						
General	Any other useful information to help Estimator						

APPENDIX II

Specimen Letters of Invitation to Tender

Dear Sirs,

New Street, C.D.A.-Phase 1

130 Houses for the Local Authority

We are presently tendering for the above project and we invite you to submit an offer for the Manufacture, Supply and Delivery of the materials as detailed on the enclosed lists and Bill of Quantities.

It is anticipated that the contract, which will be on a FIXED PRICE BASIS, will commence within the next 3 months with a contract duration of 15 months.

Your offer, based on the above, showing Trade and Cash Discounts, together with delivery periods, must reach this office not later than first post on ....197..

If you are unable to submit an offer, please advise the writer by return of post.

Yours faithfully,

....

Dear Sirs,

Invitation to Tender for Sub-Contract Work

Job: 12 Houses at Oak Place & 10 Houses & 5 Garages at Beechwood Forest, Angus

We are presently tendering as Main Contractors for the above project, and invite you to submit an estimate to us for

in accordance with enclosed trade preambles and sub-trade Bill of Quantities, which must be

returned to reach this office by first post on
............197..

Notwithstanding any standard imprinted Conditions of Sale which you are accustomed to impose, the attached Particulars and Conditions of the Main Contract will be held to prevail.

It is our intention to employ the Standard Form of Building Sub-Contract issued under the sanction of the Scottish Building Contract Committee when accepting any Sub-Contractors for this contract.

Your offer should be open for acceptance for a period of three months.

Should you wish to be excused from tendering in this instance, please advise this office immediately.

Yours faithfully,

Note: The 'Green Form' Sub-Contract is issued in England under the sanction of and approved by the National Federation of Building Trades Employers Federation of Associations of Specialists and Sub-Contractors.

The 'Pink Form' Building Sub-Contract is the Scottish equivalent of the 'Green Form' and is issued under the sanction of the Scottish Building Contract Committee.

This form is, at present, used for both domestic and nominated sub-contractors.

2 'All-in' Rate

N.B. The following is based on the Working Rules as approved by the Scottish Regional Committee and by the National Joint Council for the Building Industry. Reference should be made by the reader to any local committee agreement concerning the Working Rules. It is important to remember that within the National Working Rules there are numerous regional or area variations and additions (nationally approved) which should be ascertained from the Regional Joint Committee Secretaries or from the appropriate regional or local booklets. In view of the fact that some Regions have converted mileage to kilometres, we have shown two examples on pp. 45–52 based on the Working Rule as approved by the Northern Counties Regional Joint Committee for the Building Industry.

LABOUR COSTS

The first step in preparing estimates or in analysing prices is to determine the weekly labour costs and from these figures calculate the hourly 'Labour Rate'.

Labour Rates depend on the size and situation of the job and when applicable can include for the following:

Direct Costs

(Gross pay as defined in leaflet NP15, see later.)

(1) Weekly wage including the Basic Weekly Wage, the Joint Board Supplement and the Guaranteed Minimum Bonus Payment, plus extra payments as set out in the Working Rules. Alternatively the extra payments can be added on when costing output (see examples).
(2) Travelling Time.
(3) Overtime.
(4) Supervision.

(5) Sick Pay (pay continuing while employee is sick or otherwise absent from work).
(6) Public Holidays (amounts set aside throughout the year to be paid to him at a certain time).
(7) The 1977 Supplement.
(8) National Insurance Contributions. (*See* Note 3 at end of 'All-in Rate' on p. 52.)
(9) Annual Holidays with Pay.

Indirect Costs

(10) Common Law Insurance.
(11) CITB Levy.
(12) Travelling Expenses.
(13) Lodging allowance.
(14) Tool money.
(15) Redundancy Payment Schemes. Absenteeism etc.
(16) Tea breaks.
(17) Guaranteed week.

(1) Weekly Wages

The standard wage rates are determined by the National Joint Council for the Building Industry and from 28 June 1976 will be:

	Craftsmen £	Labourers £
Standard Basic Grade A*	37·00	31·40
Joint Board Supplement	11·00	10·20
Guaranteed Minimum Bonus Payment	4·00	3·60
	52·00	45·20

*Plus 20p in London and Liverpool.

For apprentices serving a 3 year apprenticeship entered into on or after 12 August 1974:

		% of Craft Operatives Rate	Standard Basic Rate	JBS	GMB	Total
1st year	1st 6 months	45%	£16·65	£2·25	£1·80	£20·70
	2nd 6 months	50%	£18·50	£5·50	£2·00	£26·00
2nd year	1st 6 months	60%	£22·20	£6·60	£2·40	£31·20
	2nd 6 months	75%	£27·75	£8·25	£3·00	£39·00
3rd year	1st 6 months	85%	£31·45	£9·35	£3·40	£44·20
	2nd 6 months	90%	£33·30	£9·90	£3·60	£46·80

To the above figures, from the 27 June 1977, operatives shall be entitled to a weekly supplement to be known as 'the 1977 Supplement'. This is to be 5% of each individual operatives 'total earnings' subject amongst others to the following:

(*a*) 'Total earnings' shall be the gross pay used for calculating the National Insurance contributions other than the supplement itself.
(*b*) The supplement shall in no case exceed £4 nor be less than £2.50.
(*c*) The supplement is *not* to be added to payment under the Building and Civil Engineering Industries Annual Holidays Agreement.

The settlement is to be for 12 months and no application or recommendation for a change in operatives' pay or for a change in conditions of a major character are to be considered before 26 June 1978.

The National Joint Council for the Building Industry is a body consisting of equal number of representatives from both sides of the industry (Employers and Employees) who decide on standard rates of wages and also determine conditions of employment in the Building Industry. The working rules issued by this body should be carefully studied. There is also provision in these rules for extra wages which may be paid in certain circumstances, e.g.

Carpenters required to re-use materials for concrete work –
Masons or bricklayers engaged in re-dressing and cutting for indents in stairs –
Slaters employed in the fixing and repair of concrete sand-faced roofing tiles –
Labourer employed in and actually responsible for

operating a concrete mixer up to and including 10/7 capacity or mortar pan or barrow hoist (to apply to one man per machine only).

Over and above the weekly wages certain tradesmen are paid tool allowances for providing and maintaining their own tools.

(2) Travelling Time

When workmen have to travel to reach the site travelling time shall be paid as follows:

over 5 miles and under 10 miles – $\frac{1}{4}$ hour per day
over 10 miles and under 15 miles – $\frac{1}{2}$ hour per day
over 15 miles and under 20 miles – $\frac{3}{4}$ hour per day
over 20 miles and under 25 miles – 1 hour per day

and an additional $\frac{1}{4}$ hour per day for every 5 miles or part thereof in excess of the above.

The distance is measured from the nearest transport route in the area from which the men may travel. Travelling time is only paid one way unless the distance is over 25 miles in which case $\frac{1}{4}$ hour per day is paid for every 5 miles or part thereof in excess of this figure on the return journey.

Example

For a distance of 35 miles

outward travelling time – $1\frac{1}{2}$ hours per day
inward travelling time – $\frac{1}{2}$ hour per day

Total 2 hours per day

In the case of men travelling to the shop or yard the maximum daily allowance for travelling time will be 10 miles (i.e. $\frac{1}{4}$ hour). If the distance exceeds this, the man should pay for any additional cost.

(3) Overtime (W.R. 4)

The normal working week is 40 hours on the basis of five 8 hour days Monday to Friday. The hours of work shall be between 7.30 a.m. and 5.00 p.m.

Overtime shall be paid at the following rates:

> From Monday to Friday from agreed stopping time until 10.30 p.m., time and a half, afterwards until starting time next morning, and from 10.30 p.m. on Friday until 8.00 a.m. on Saturday, double time. Time worked on Saturday until 12 noon, time and a half, afterwards until starting time on Monday morning, double time.

(4) Supervision

Trade charge-hands and gangers are paid above the standard rate for craftsmen or for labourers as the case may be. The additional cost of this must be calculated and divided by the total number of men employed on the job.

e.g. Squad required is 10 men, one of whom is paid 10p per hour extra to act as chargehand.

The allowances for supervision would be calculated as follows:

Normal working week	40 hours
Travelling time say	5 hours
∴ Total time	45 hours

∴ Cost of chargehand = 45 hours @ 10p = £4·50
Total number of men employed = 10

$$\therefore \text{Average cost per man per week} = \frac{£4{\cdot}50}{10} = \underline{\underline{45p}}$$

If overtime was worked this would also have to be allowed for in the calculation of the cost of supervision.

The additional rates paid to chargehands or gangers vary from 8p–15p per hour above the craftsmen's or labourer's rate.

(5) Sick Pay (W.R. 8)

An operative absent from work on account of sickness shall be paid £1·50 per working day subject to certain conditions one of which is payments are limited to paying that amount by which earnings related supplement paid by the DHSS falls short of £1·50 per day. This has been allowed for in the 'All-in-rate' as occurring 5 days per annum.

(6) Public Holidays

The Public Holiday Pay Scheme ceased to operate on April 1971 the firms making their own provision for financing wage payments on public holidays. This change brought this allowance within the meaning of gross pay for National Insurance contributions whereas the Annual Holiday Pay Scheme does not (*see* leaflet NP15).

(7) The 1977 Supplement

The 1977 Supplement is to be paid first for the week beginning 27 June 1977. It is applied by ascertaining total payments from which National Insurance contributions would be deducted if there were no 1977 Supplement. Take 5% of these total payments. Reduce the result to £4 where it exceeds that figure or make it up to the £2·50 minimum if necessary. The 1977 Supplement is added to the total payments with tax and National Insurance deductions being calculated on the new total.

Sick pay and payment for public holidays should be included in the total payments for calculation of the 1977 Supplement. Non-taxable items are not to be included in the calculation (e.g. fare and lodging allowances) neither are tool allowances or sums paid in encashment of holiday stamps under the Holidays-with-Pay Scheme.

(8) National Insurance Contributions
(*See* Note 3 at end of 'All-in Rate' on p. 52)

The graduated pension scheme has now been abandoned. On the 6 April 1975 the Social Security Act 1973 came into operation. From that date flat-rate and graduated contributions ceased to be payable for employees and were replaced by wholly earnings-related contributions. Stamp cards for employees were abolished at the same time and the earnings-related national insurance contributions are now collected with income tax under the PAYE procedure.

Liability for the contributions of both employers and employees is limited by upper (£105 per week) and lower (£15 per week) earnings limits.

National Insurance contributions are always calculated on gross pay (*see* DHSS leaflet NP15, 'Employers' Guide to National Insurance Contributions).

An employer may calculate contributions either:

(*a*) exactly by applying the prescribed percentages; or
(*b*) by reference to contribution tables supplied by the Department of Health and Social Security (National Insurance contribution tables CF391);
(*c*) in many instances wages are now made up by computer which necessitates the use of percentages and the employer would contribute as follows for a Tradesman's gross wage of £60·46 and a Labourer's gross wage of £52·46:

Tradesman: $10\frac{3}{4}$% on £60·46 = £6·50
Labourer: $10\frac{3}{4}$% on £52·46 = £5·64.

Each calculation is rounded to the nearest penny.

Check this with the weekly Table A in CF391 and the amounts will be found to be £6·48 and £5·62 respectively. This is because the table is produced in 50p increases in gross pay and if the exact gross pay figure is not shown in the table, the next smaller figure down is used.

The amounts shown in the Ministry leaflet will be used in the examples.

(9) Holidays with Pay

The employer must buy a stamp to cover holiday pay, the value of the stamp being refunded at holiday time to cover wages. The weekly contributions are £4·05 to be increased by 15p per week (juveniles proportionately) with effect from 1 August 1977, with this amount being augmented by up to a further 5p per week from the funds of the Building and Civil Engineering Holidays Scheme Management Co.

(10) Common Law Insurance

This covers the contractor for all claims at common law. The premiums vary but are based on the amount of wages paid. A normal quotation for an average job is £2·00 per £100 and this will be used in all calculations.

(11) The Construction Industry Training Board

The CITB was established by the Government under the SI (The Industrial Training (Construction) Board Order 1964 SI 1974 No 1079) and became effective on 21 July 1964. The purpose of the board is to back training in the industry and to enable it to do this a levy is applied on firms. Although allowed for in our calculations some firms include for this cost in overheads.

The levy is based on the type of man employed on a varying scale, i.e.

Cost per Annum 1977/78

Tradesmen (Bricklayers and Joiners)	£20
Electrician, Plumbers, Engineers	£35
Labourer	£ 5

(12) Travelling Expenses

Travelling expenses to and from the job by public transport are paid by the employer less 4/5ths of the man's hourly rate.

e.g. Men travel to a job at the return daily fare of 50p
The employer's contribution is calculated as follows:
Daily cost 50p Weekly cost 50 x 5 = 250p
Employers contribution =
250p less 4/5ths of 92½p (in the case of tradesmen)
or 250p less 4/5ths of 78½p (in the case of labourers)

The men may be transported to the site by the employers lorry in which case the cost per workman must be calculated and added to the rate.

(13) Lodging Allowances

This is given in the Working Rules at an allowance of £26·25 per week from commencement.

All men sent to jobs beyond 4 miles and up to 20 miles from the agreed boundary shall be allowed train fares or be conveyed to and from the job every week; if more than 20 miles and up to 40 miles, every fortnight, from 40 to 80 miles, once every month, over 80 miles mutual arrangements to be made.

(14) Tool Money

Because tool allowances are regarded as maintenance and upkeep to tools provided by the operatives they are not deemed to be wage payments.

(15) Redundancy Payments Scheme

See leaflet (tenth revision effective from 1 June 1976) issued by the Department of Employment giving general guidance to employees and employers about their rights and obligations under the Redundancy Payments Acts 1965 and 1969 with references to related Acts including Contracts of Employment Act 1972, Trade Union and Labour Relations Act 1974 and Employment Protection Act 1975.

(16) Tea Breaks

It is normal to allow for a tea break of 10 minutes per day. In the example we will take a % of $2\frac{1}{2}$ to cover for this.

(17) Guaranteed Week

The working rules state that the minimum weekly wage shall be 40 hours per week subject to certain provisions, which must be paid when the men cannot work due to inclement weather, plant breakdown, non-arrival of materials, etc. This is known as the 'guaranteed weekly minimum'. It is almost impossible to make an accurate allowance for guaranteed time and, if asked for, can only be covered by adding a percentage to the Labour Rate. This percentage obviously varies with the type of job, the time of year, etc. In most examinations the Candidate does not have to make this allowance and on the few occasions which it has been asked the percentage has been given. Some suggested percentages and the ones which will be used in examples are:

Bricklayers and Labourers	7½%–10%
Joiners	2½%–5%
Slaters	7½%–10%
Plasterers	2½%–5%
Plumbers	nil or 2½%

General Note

Extra payments required by the terms of the Working Rule Agreement for:

(*a*) Discomfort, Inconvenience or Risk;
(*b*) Continuous Extra Skill or Responsibility;
(*c*) Intermittent Responsibility;

are not included in the preparation of the 'all-in' rate. They are added to the 'all-in' rate, where required, in the computation of the individual rates. (See note at beginning of chapter.)

Example 1

CIRCUMSTANCES: Contractor wishes to tender for a contract 7 miles from yard (Working Rule expressed in imperial terms) which because of the nature of the job there will be no overtime. Average of 6 tradesmen and 4 labourers plus a foreman who will be paid 10p per hour extra. Bus fare is £2·00 per week. Guaranteed time is taken as 7½%.

Description	Tradesman's Unit Rate	Labourer's Unit Rate	Tradesman	Labourer
1. Basic Weekly Wage Grade A and Scotland	92½p per hour	78½p per hour	37·00	31·40
(*a*) Plus Rate (Dirty money etc.)				
(*b*) Joint Board Supplement			11·00	10·20
(*c*) Guaranteed Minimum Bonus Payment			4·00	3·60
		£	52·00	45·20
2. Overtime				
		C/f £	52·00	45·20

Description	Tradesman's Unit Rate	Labourer's Unit Rate	Tradesman	Labourer
B/f £			52·00	45·20
3. <u>Travelling Time</u> 5 to 10 miles = $\frac{1}{4}$ hour per day 5 days x $\frac{1}{4}$ hour = $1\frac{1}{4}$ hour per week T = $1\frac{1}{4}$ @	$92\frac{1}{2}$p		1·16	
L = $1\frac{1}{4}$ @		$78\frac{1}{2}$		0·98
4. <u>Supervision</u> 40 hours Basic $1\frac{1}{4}$ hours Travelling Time $41\frac{1}{4}$ hours @ 10p ÷ 11 men			0·38	0·38
£			53·54	46·56
5. <u>Sick Pay</u> 5 days per annum 5 x £1·50 = £7·50 ÷ 48 weeks (48 weeks based on 3 weeks annual and 7 days public holidays)			$0·15\frac{1}{2}$	$0·15\frac{1}{2}$
6. <u>Public Holidays</u> 7 days <u>Tradesman</u> Weekly rate = £52·00 plus Supervision $\frac{10p \times 40\ hrs}{11\ men}$ = £ 0·36 £52·36 Hourly Rate = $\frac{£52·36}{40}$ $\left(7\ days \times 8\ hours \times \frac{£52·36}{40}\right) \div 48$ working weeks			1·53	
<u>Labourer</u> £45·20 plus £0·36 = £45·56 $\left(7\ days \times 8\ hours \times \frac{£45·56}{40}\right) \div 48$ working weeks				1·33
			$55·22\frac{1}{2}$	$48·04\frac{1}{2}$
7. <u>The 1977 Supplement (from 27/6/77)</u> 5% (£4 maximum £2·50 minimum)			$2·76\frac{1}{2}$	2·50
Adjustments for production bonus should be considered when applying the Supplement, i.e. £4 maximum				
C/f £			57·99	$50·54\frac{1}{2}$

Description	Tradesman	Labourer
B/f £	57·99	50·54½
8. <u>National Insurance Contributions</u>		
Standard-Rate Contributions		
Weekly Table A — T 6·21, L 5·46	6·21	5·46
Amount of Annual Holidays Excluded (*see* leaflet NP15 issued by DHSS)		
£	64·20	56·00½
9. <u>Annual Holidays with Pay</u>	4·20	4·20
£	68·40	60·20½
10. <u>Employer's Liability and Third Party Insurance</u>		
2% (£2·00 per £100) on £68·40 and £60·20½	1·37	1·20
11. <u>CITB Levy</u>		
Tradesman £20·00 ÷ 48	0·41½	
Labourer £ 5·00 ÷ 48		0·10½
12. <u>Travelling Expenses</u>		
Weekly Cost $- \frac{4}{5} \times \frac{\text{Weekly Rate}}{40}$		
Tradesman £2·00 $- \frac{4}{5} \times$ 92½p)	1·26	
Labourer £2·00 $- \frac{4}{5} \times$ 78½p)		1·37
13. <u>Lodging Allowance</u>		
NIL		
14. <u>Tool Money</u> (25p per week is the allowance for most trades with some exceptions; e.g. carpenter 50p, banker mason 38p)	0·25	–
C/f £	71·69½	62·88

Description	Tradesman	Labourer
B/f £	71·69½	62·88
15. Redundancy Pay; Absenteeism; Other Marginal Costs		
1½%	1·07½	0·94
16. Tea Breaks (5 days x 12 mins) = 1 hour in 40 hours		
2½%	1·79	1·57
17. Guaranteed Time		
7½%	5·38	4·71
Gross Weekly Total £	79·94	70·10
For 'All-in' Hourly Rate ÷ by hours worked (i.e. Basic plus Overtime) 40 £	2·00	1·75

Example 2

CIRCUMSTANCES: Contract is 19 miles from yard to site. Men transported by contractor's own vehicle which is used on site and costs £4·00 per hour including driver. Assumed lorry averages 20 m.p.h. Normal site requirements 16 men per day including foreman paid 12p per hour extra for supervision. Otherwise as Example 1. (Cost of lorry added to material costs, etc. at other times.)

Description	Tradesman's Unit Rate	Labourer's Unit Rate	Tradesman	Labourer
1. Basic Weekly Wage				
Grade A and Scotland	$92\frac{1}{2}$p per hour	$78\frac{1}{2}$p per hour	37·00	31·40
(a) Plus Rate (Dirty money etc.)				
(b) Joint Board Supplement			11·00	10·20
(c) Guaranteed Minimum Bonus Payment			4·00	3·60
		£	52·00	45·20
2. Overtime				
NIL				
3. Travelling Time				
15 to 20 miles = $\frac{3}{4}$ hours per day				
5 days x $\frac{3}{4}$ hour = $3\frac{3}{4}$ hours				
T = $3\frac{3}{4}$ hours @	$92\frac{1}{2}$p		3·47	
L = $3\frac{3}{4}$ hours @		$78\frac{1}{2}$p		2·94
4. Supervision 40 hours Basic				
$3\frac{3}{4}$ hours Travelling Time				
$43\frac{3}{4}$ hours @ 12p ÷ 16 men			0·33	0·33
		£	55·80	48·47
5. Sick Pay 5 days per annum				
5 x £1·50 = £7·50 ÷ 48 weeks (48 weeks based on 3 weeks annual and 7 days public holidays)			$0{\cdot}15\frac{1}{2}$	$0{\cdot}15\frac{1}{2}$
		C/f £	$55{\cdot}95\frac{1}{2}$	$48{\cdot}62\frac{1}{2}$

Description	Tradesman	Labourer
B/f £	55·95½	48·62½
6. <u>Public Holidays</u>		
<u>Tradesman</u>		
Weekly Rate = £52·00		
plus Supervision $\frac{12\text{p} \times 40 \text{ hrs}}{16 \text{ men}}$ = £ 0·30		
£52·30		
Hourly Rate = $\frac{£52{\cdot}30}{40}$		
$\left(7 \text{ days} \times 8 \text{ hours} \times \frac{£52{\cdot}30}{40}\right) \div 48$ working weeks	1·52½	
<u>Labourer</u>		
£45·20 plus £0·33 = £45·50		
$\left(7 \text{ days} \times 8 \text{ hours} \times \frac{£45{\cdot}40}{40}\right) \div 48$ working weeks		1·33
7. <u>The 1977 Supplement</u>	57·58	49·95½
5% (£4 maximum £2·50 minimum)	2·88	2·50
	60·46	52·45½
8. <u>National Insurance Contributions</u>		
Standard Rate Contributions		
Weekly Table A — T 6·48, L 5·62	6·48	5·62
9. <u>Annual Holidays with Pay</u>	4·20	4·20
	71·14	62·27½
10. <u>Employer's Liability and Third Party Insurance</u>		
2% (£2·00 per £100) on T = £71·14	1·42	
L = £62·27½		1·25
C/f £	72·56	63·52½

Description	Tradesman	Labourer
B/f £	72·56	63·52½
11. CITB Levy		
T = £20·00 ÷ 48	0·41½	
L = £ 5·00 ÷ 48		0·10½
12. Travelling Expenses		
Weekly Cost – $\left(\frac{4}{5} \times \frac{\text{Weekly Rate}}{40}\right)$ or Average Cost		
20 m.p.h. say 2 hours per day there and back		
5 days x 2 hours = 10 hours @ £4·00 per hour		
= £40·00		
£40·00 ÷ 16	2·50	2·50
£	75·47½	66·13
13. Lodging Allowance	—	—
14. Tool Money	0·25	—
	75·72½	66·13
15. Redundancy Pay; Absenteeism; Other Marginal Costs		
1½%	1·13½	0·99
16. Tea Breaks		
2½%	1·89	1·65
17. Guaranteed Time		
7½%	5·68	4·96
Gross Weekly Total £	84·43	73·73
For 'All-in' Hourly Rate ÷ by hours worked (i.e. Basic plus Overtime) 40 £	2·11	1·84

Note: Difference in time between item 3, Travelling Time of ¾ hour and item 12, Travelling Expenses of 1 hour there and the same back is because Travelling Time is an *allowance* whereas time taken by the lorry is *actual* time.

Example 3

CIRCUMSTANCES: Carpenters sent to a country job 36 miles from contractor's yard. (No tradesmen available in town where work is to be carried out.) Squad is composed of 6 men and overtime is to be 3 hours per day (agreed with Trade Union) Monday to Thursday. Normal working week is 5 days. Duration of work estimated to be 27 weeks. Return fare £2·00. Foreman paid 12 p per hour extra. Labourer to be engaged in town to tidy up, etc.

Description	Tradesman's Unit Rate	Labourer's Unit Rate	Tradesman	Labourer
1. <u>Basic Weekly Wage</u>				
Grade A and Scotland	$92\frac{1}{2}$p per hour	$78\frac{1}{2}$p per hour	37·00	31·40
(*a*) Plus Rate (Dirty money etc.)				
(*b*) Joint Board Supplement			11·00	10·20
(*c*) Guaranteed Minimum Bonus Payment			4·00	3·60
		£	52·00	45·20
2. <u>Overtime</u>				
3 hours x 4 days = 12 hours				
12 hours x Time and a half = 18 hours				
T = 18 hours @	$92\frac{1}{2}$p		16·65	
L = 18 hours @		$78\frac{1}{2}$p		14·13
		C/f £	68·65	59·33

Description	Tradesman's Unit Rate	Labourer's Unit Rate	Tradesman	Labourer
B/f £			68·65	59·33
3. Travelling Time 35 to 40 miles = 1¾ hours plus 1¾ − 1 hr = ¾ hour 2½ hours 2½ hours per fortnight ∴ 1¼ hours per week @	92½p		1·16	
Coming back from job at finish 1¾ plus for last week going 1¾ 3½ ÷ 27 @ (if 26 weeks only 1 hour as 1 hour has been ddt from 1¾)	92½p		0·12	
4. Supervision 40 hours Basic 18 hours Overtime 1¼ hours Travelling Time $\frac{7}{54}$ hours Travelling in last week 59·38 @ 12p ÷ 7			1·02	1·02
			70·95	60·35
5. Sick Pay			0·15½	0·15½
6. Public Holidays Tradesman Weekly Rate £52·00 plus Supervision $\frac{12p \times 40 \text{ hrs}}{7 \text{ men}}$ £ 0·69 £52·69 Hourly Rate = $\frac{£52{\cdot}69}{40}$ $\left(7 \text{ days} \times 8 \text{ hours} \times \frac{£52{\cdot}69}{40}\right) \div 48$ working weeks			1·53½	
Labourer £45·20 plus 0·69 = £45·89 $\left(7 \text{ days} \times 8 \text{ hours} \times \frac{£45{\cdot}89}{40}\right) \div 48$ working weeks				1·34
C/f £			72·64	61·84½

Description	Tradesman	Labourer
B/f £	72·64	61·84½
7. <u>The 1977 Supplement</u>		
5% (£4 maximum £2·50 minimum)	3·63	3·09
£	76·27	64·93½
8. <u>National Insurance Contributions</u>		
Standard Rate Contributions: T L Weekly Table A 8·20 6·96	8·20	6·96
9. <u>Annual Holidays with Pay</u>	4·20	4·20
£	88·67	76·09½
10. <u>Employer's Liability and Third Party Insurance</u>		
2% (£2·00 per £100) on T = 88·67	1·77	
L = 76·09½		1·52
11. <u>CITB Levy</u>		
T = £20·00 ÷ 48	0·41½	
L = £ 5·00 ÷ 48		0·10½
12. <u>Travelling Expenses</u>		
Weekly Cost – $\frac{4}{5} \times \frac{\text{Weekly Rate}}{40}$ or Average Cost		
£2·00 every 2 weeks ∴ £1 per week	1·00	
plus £2·00 in 27 weeks for return fare in last week	0·07½	
£	91·93	77·72
13. <u>Lodging Allowance</u>	26·25	—
14. <u>Tool Money</u>	0·50	
C/f £	118·68	77·72

Description	Tradesman	Labourer
B/f £	118·68	77·72
15. Redundency Pay; Absenteeism; Other Marginal Costs		
1½%	1·78	1·17
16. Tea Breaks		
2½%	2·97	1·94½
17. Guaranteed Time		
2½%	2·97	1·94½
Gross Weekly Total £	126·40	82·78
For 'All-in' Hourly Rate ÷ by hours worked (i.e. basic plus Overtime) 52 £	2·43	1·59

Note: Travelling time can be by negotiation under such circumstances as described above but ¼ hour for every 5 miles is a good guide.

Example 4

CIRCUMSTANCES: Build up an 'All-in' Rate for work 14 miles from the office; the men (2 tradesmen and 1 labourer) will travel by public transport, the return fare being 65p per day. Allow 8 hours overtime. The normal working week is 5 days. One tradesman is paid 10p per hour extra. 10% Guaranteed Time.

Description	Tradesman's Unit Rate	Labourer's Unit Rate	Tradesman	Labourer
1. Basic Weekly Wage				
Grade A and Scotland	92½p per hour	78½p per hour	37·00	31·40
(*a*) Plus Rate (Dirty money etc.) (*b*) Joint Board Supplement			11·00	10·20
(*c*) Guaranteed Minimum Bonus Payment			4·00	3·60
		£	52·00	45·20
2. Overtime				
8 hours × 1½ = 12				
T = 12 @	92½p		11·10	
L = 12 @		78½p		9·42
3. Travelling Time				
½ hour per day × 5 days = 2½ hours	92½p	78½p	2·44	1·96
4. Supervision 40 hours Basic; 12 hours Overtime; 2½ hours Travelling Time; 54½ × 10p = 545 ÷ 3			1·82	1·82
		C/f £	67·36	58·40

Description		Tradesman	Labourer
	B/f £	67·36	58·40
5. Sick Pay		0·15½	0·15½
6. Public Holidays			
Tradesman Weekly rate	£52·00		
plus Supervision $\frac{10p \times 40 \text{ hrs}}{3 \text{ men}}$	£ 1·33		
	£53·33		
$\left(7 \text{ days} \times 8 \text{ hours} \times \frac{£53{\cdot}33}{40}\right) \div 48$ working weeks		1·55½	
Labourer £45·20 plus £1·33 = £46·43			
$\left(7 \text{ days} \times 8 \text{ hours} \times \frac{£46{\cdot}53}{40}\right) \div 48$ working weeks			1·35½
		69·07	59·91
7. The 1977 Supplement			
5% (£4 maximum £2·50 minimum)		3·45	3·00
	£	72·52	62·91
8. National Insurance Contributions			
Standard Rate Contributions			
T L			
Weekly Table A 7·87 6·75		7·82	6·75
9. Annual Holidays with Pay		4·20	4·20
	C/f £	84·54	73·86

Description	Tradesman	Labourer
B/f £	84·54	73·86
10. Employer's Liability and Third Party Insurance		
2% (£2·00 per £100) on T = £84·54	1·69	
L = £73·86		$1{\cdot}47\frac{1}{2}$
11. CITB Levy		
T = £20·00 ÷ 48	$0{\cdot}41\frac{1}{2}$	
L = £ 5·00 ÷ 48		$0{\cdot}10\frac{1}{2}$
12. Travelling Expenses		
Weekly Cost $-\left(\frac{4}{5}\times\frac{\text{Weekly Rate}}{40}\right)$ or Average Cost		
Tradesman = £3·25 $-\left(\frac{4}{5}\times\frac{£37}{40}\right)$	2·51	
Labourer = £3·25 $-\left(\frac{4}{5}\times\frac{£31{\cdot}40}{40}\right)$		2·62
13. Lodging Allowance	–	–
14. Tool Money	0·25	
£	$89{\cdot}40\frac{1}{2}$	78·06
15. Redundancy Pay; Absenteeism, Other Marginal Costs		
$1\frac{1}{2}$%	1·35	1·17
16. Tea Breaks		
$2\frac{1}{2}$%	2·25	1·95
17. Guaranteed Time		
10%	9·01	7·81
Gross Weekly Total £	$102{\cdot}01\frac{1}{2}$	88·99
For 'All-in' Hourly Rate ÷ by hours worked 48 £	2·13	1·85

Example 5

CIRCUMSTANCES: Men travel 9 miles, fare 50p per day. 1½ hours overtime Monday to Thursday. 5 day week. Foreman included in 12 men @ 10p per hour extra. 10% Guaranteed Time.

Description	Tradesman's Unit Rate	Labourer's Unit Rate	Tradesman	Labourer
1. <u>Basic Weekly Wage</u>				
Grade A and Scotland	92½p per hour	78½p per hour	37·00	31·40
(*a*) Plus Rate (Dirty money etc.)				
(*b*) Joint Board Supplement			11·00	10·20
(*c*) Guaranteed Minimum Bonus Payment			4·00	3·60
		£	52·00	45·20
2. <u>Overtime</u>				
4 x 1½ = 6 hours plus 3 hours				
9 hours @	92½p	78½p	8·33	7·07
3. <u>Travelling Time</u>				
¼ hour per day x 5 = 1¼ hours @	92½p	78½p	1·15	0·98
4. <u>Supervision</u> 40 hours Basic 9 hours Overtime 1¼ hours Travelling Time				
50¼ hours x 10p ÷ 12			0·42	0·42
		£	61·90	53·67
5. <u>Sick pay</u>			0·15½	0·15½
		C/f £	62·05½	53·82½

Description	Tradesman	Labourer
B/f £	62·05½	53·82½
6. Public Holidays		
Tradesman £52·00 plus $\frac{10\text{p} \times 40 \text{ hrs}}{12}$ = £52·34		
$\left(7 \text{ days} \times 8 \text{ hours} \times \frac{£52{\cdot}34}{40}\right) \div 48$ working weeks	1·52½	
Labourer £45·20 plus £0·34 = £45·54		
$\left(7 \text{ days} \times 8 \text{ hours} \times \frac{£45{\cdot}54}{40}\right) \div 48$ working weeks		1·32
£	63·58	55·14½
7. The 1977 Supplement		
5% (£4 maximum £2·50 minimum)	3·18	2·76
8. National Insurance Contributions £	66·76	57·90½
Standard Rate Contributions: T / L — Weekly Table A 7·18 6·21	7·18	6·21
9. Annual Holidays with Pay	4·20	4·20
£	78·14	68·31½
10. Employer's Liability and Third Party Insurance		
2% (£2·00 per £100) on T = £79·29	1·56	
L = £68·31½		1·36½
11. CITB Levy		
Tradesman £20·00 ÷ 48	0·41½	
Labourer £ 5·00 ÷ 48		0·10½
12. Travelling Expenses		
Weekly Cost $-\left(\frac{4}{5} \times \frac{\text{Weekly Rate}}{40}\right)$ or Average Cost		
Tradesman = £2·50 $-\left(\frac{4}{5} \times \frac{£37}{40}\right)$ =	1·76	
Labourer = £2·50 $-\left(\frac{4}{5} \times \frac{£31{\cdot}40}{40}\right)$ =		1·87
C/f £	81·87½	71·65½

Description	Tradesman	Labourer
Bf/ £	81·87½	71·65½
13. Lodging Allowance	–	–
14. Tool Money	0·25	
£	82·12½	71·65½
15. Redundancy Pay; Absenteeism, Other Marginal Costs		
1½%	1·23	1·07½
16. Tea Breaks		
2½%	2·05	1·79
17. Guaranteed Time		
10%	8·21	7·16½
Gross Weekly Total £	93·61½	81·68½
For 'All-in' Hourly Rate ÷ by hours worked 46 £	2·04	1·78

Note: The rates to be used in all following examples will be £2·20 for a tradesman and £1·90 for a labourer.

Based on the Working Rules as approved by the Northern Counties Regional Joint Committee for the Building Industry

Example 1

CIRCUMSTANCES: Contract is 30 km from yard to site. Men transported by contractor's own vehicle which is used on site and costs £4·00 per hour including driver. Assumed lorry averages 20 m.p.h. Normal site requirements 16 men per day including foreman paid 12p per hour extra for supervision. Otherwise as Example 1. (Cost of lorry added to material costs, etc. at other times.)

Description	Tradesman's Unit Rate	Labourer's Unit Rate	Tradesman	Labourer
1. Basic Weekly Wage				
Grade A and Scotland	$92\frac{1}{2}$p per hour	$78\frac{1}{2}$p per hour	37·00	31·40
(*a*) Plus Rate (Dirty money etc.)				
(*b*) Joint Board Supplement			11·00	10·20
(*c*) Guaranteed Minimum Bonus Payment			4·00	3·60
		£	52·00	45·20
2. Overtime				
NIL				
3. Travelling Time W.R. 6.A.2 5 x 5 x (30 – 6)			6·00	6·00
4. Supervision				
40 hours Basic $3\frac{3}{4}$ hours Travelling Time				
$43\frac{3}{4}$ hours @ 12p ÷ 16 men			0·33	0·33
		£	58·33	51·53
5. Sick Pay 5 days per annum 5 x £1·50 = £7·50 ÷ 48 weeks (48 weeks based on 3 weeks annual and 7 days public holidays)			$0·15\frac{1}{2}$	$0·15\frac{1}{2}$
		£	$58·48\frac{1}{2}$	$51·68\frac{1}{2}$
6. Public Holidays Tradesman Weekly Rate = £52·00 plus Supervision $\frac{12\text{p} \times 40\text{ hours}}{16\text{ men}}$ = £ 0·30 £52·30 Hourly Rate = $\frac{£52·30}{40}$ $\left(7\text{ days} \times 8\text{ hours} \times \frac{£52·30}{40}\right) \div 48$ working weeks			$1·52\frac{1}{2}$	
		C/fd £		

Description		Tradesman	Labourer
	B/fd £		
Labourer £45·20 plus £0·30 = £45·50 $\left(7 \text{ days} \times 8 \text{ hours} \times \frac{£45{\cdot}50}{40}\right) \div 48$ working weeks			1·33
	£	60·01	53·01½
7. The 1977 Supplement 5% (£4 maximum £2·50 minimum)		3·00	2·65
	£	63·01	55·66½
8. National Insurance Contributions Standard Rate Contributions: T L Weekly Table A 6·80 5·99		6·80	5·99
9. Annual Holidays with Pay		4·20	4·20
	£	74·01	65·85½
10. Employer's Liability and Third Party Insurance 2% (£2·00 per £100) on T = £74·01		1·48	
L = £65·85½			1·31½
	£	75·49	67·17
11. CITB Levy T = £20·00 ÷ 48		0·41½	
L = £ 5·00 ÷ 48			0·10½
12. Travelling Expenses Average Cost 20 m.p.h. say 2 hours per day there and back 5 days x 2 hours = 10 hours @ £4·00 per hour = £40·00 £40·00 ÷ 16	£	2·50	2·50
	£	78·40½	69·77½
13. Lodging Allowance			
14. Tool Money		0·25	
	£	78·65½	69·77½
15. Redundancy Pay; Absenteeism; Other Marginal Costs 1½%		1·18	1·04½
16. Tea Breaks 2½%		1·96½	1·74
	C/fd £	81·80	72·56

Description		Tradesman	Labourer
	B/fd £	81·80	72·56
17. Guaranteed Time $7\frac{1}{2}$%		5·89½	5·22
	Gross Weekly Total £	87·69½	77·78
For 'All-in' Rate ÷ by hours worked (i.e. Basic plus Overtime)	÷ 40	£ 2·19	£ 1·94

Example 2

CIRCUMSTANCES: Carpenters sent to a country job 58 km from contractor's yard. (No tradesmen available in town where work is to be carried out.) Squad is composed of 6 men and overtime is to be 3 hours per day (agreed with Trade Union) Monday to Thursday. Normal working week is 5 days. Duration of work estimated to be 27 weeks. Return fare £2·00. Foreman paid 12p per hour extra. Labourer to be engaged in town to tidy up, etc.

Description	Tradesman's Unit Rate	Labourer's Unit Rate	Tradesman	Labourer
1. Basic Weekly Wage Grade A and Scotland	$92\frac{1}{2}$p per hour	$78\frac{1}{2}$p per hour	37·00	31·40
(*a*) Plus Rate (Dirty money etc.) (*b*) Joint Board Supplement			11·00	10·20
(*c*) Guaranteed Minimum Bonus Payment			4·00	3.60
			52·00	45·20
2. Overtime 3 hours x 4 days = 12 = 12 hours x Time and a half = 18 hours T = 18 hours @	$92\frac{1}{2}$p		16·65	
L = 18 hours @		$78\frac{1}{2}$p		14·13
3. Travelling Time 55 to 60 km = $1\frac{3}{4}$ = $1\frac{3}{4}$ hours plus $1\frac{3}{4}$ return = $1\frac{3}{4}$ hours $3\frac{1}{2}$ hours $3\frac{1}{2}$ hours per fortnight ∴ $1\frac{3}{4}$ hours per week @	$92\frac{1}{2}$p		1·62	
Coming back from job at finish $1\frac{3}{4}$ plus for last week going $1\frac{3}{4}$ $3\frac{1}{2}$ ÷ 27 @	$92\frac{1}{2}$p		0·12	
		C/fd £	70·39	59·33

Description	Tradesman	Labourer
B/fd £	70·39	59·33
4. Supervision 40 hours Basic 18 hours Overtime $1\frac{3}{4}$ hours Travelling Time $\frac{7}{54}$ hours Travelling in last week 59·88 @ 12p ÷ 7	1·03	1·03
	£ 71·42	60·36
5. Sick Pay	$0{\cdot}15\frac{1}{2}$	$0{\cdot}15\frac{1}{2}$
6. Public Holidays Tradesman Weekly Rate £52·00 plus supervision 12p x 40 hours / 7 men £ 0·69 £52·69 Hourly Rate = $\frac{£52{\cdot}69}{40}$ $\left(7 \text{ days} \times 8 \text{ hours} \times \frac{£52{\cdot}69}{40}\right) \div 48$ working weeks	$1{\cdot}53\frac{1}{2}$	
Labourer £45·20 plus £0·69 = £45·89 $\left(7 \text{ days} \times 8 \text{ hours} \times \frac{£45{\cdot}89}{40}\right) \div 48$ working weeks		1·34
£	73·11	$61{\cdot}85\frac{1}{2}$
7. The 1977 Supplement 5% (£4 maximum £2·50 minimum)	$3{\cdot}65\frac{1}{2}$	3·09
£	$76{\cdot}76\frac{1}{2}$	$64{\cdot}94\frac{1}{2}$
8. National Insurance Contributions Standard Rate Contributions: T L Weekly Table A 8·25 6·96	8·25	6·96
9. Annual Holidays with Pay	4·20	4·20
£	$89{\cdot}21\frac{1}{2}$	$76{\cdot}10\frac{1}{2}$
10. Employer's Liability and Third Party Insurance 2% (£2·00 per £100) on T = $£89{\cdot}21\frac{1}{2}$ L = $£76{\cdot}10\frac{1}{2}$	$1{\cdot}78\frac{1}{2}$	1·52
C/fd £	91·00	$77{\cdot}62\frac{1}{2}$

Description	Tradesman	Labourer
B/fd £	91·00	77·62½
11. CITB Levy		
T = £20·00 ÷ 48	0·41½	
L = £ 5·00 ÷ 48		0·10½
12. Travelling Expenses		
Average Cost		
£2·00 every 2 weeks ∴ £1 per week	1·00	
plus £2·00 in 27 weeks for return fare in last week	0·07½	
£	92·49	77·73
13. Lodging Allowance		
£3·75/night 11 nights every 2 weeks x 13		
plus 4 nights last week		
11 x 13 x 3·75 = 536·25		
4 x 3·75 = 15·00		
551·25 ÷ 27 = 20·42	20·42	
14. Tool Money	0·50	
£	113·41	77·73
15. Redundancy Pay: Absenteeism; Other Marginal Costs		
1½%	1·70	1·17
16. Tea Breaks		
2½%	2·83½	1·94½
17. Guaranteed Time		
2½%	2·83½	1·94½
Gross Weekly Total £	£ 120·78	82·79
For 'All-in' Hourly Rate ÷ by hours worked (i.e. Basic plus Overtime) 52 £	£ 2·33	1·59

Note 1: Travelling time can be by negotiation under such circumstances as described above but ¼ hour for every 8 km is a good guide.

Note 2: The interpretation of payment of Lodging Allowance is as defined by the National Federation of Building Trade Employees and the Trade Unions for the Northern Counties Regional Joint Committee for the Building Industry; i.e. on a nightly basis.

Note 3: At the time of going to print, National Insurance Contributions were based on earnings-related contributions required by the Social Security Act, 1973, and came into operation on 6 April, 1975.

From April, 1978, a new earnings-related pension scheme has been introduced. This new scheme has superseded the scheme introduced in 1975 and which introduced earnings-related contributions. From April, 1978, the Employer (not contracted out) is required to pay 10% of the employee's total earnings plus 2% under the National Insurance Surcharge Act, 1976, which gives a gross total of 12% of the employee's earnings.

Therefore at the 'All-in Rate', section 8, under the heading National Insurance Contributions the contribution of 10¾% should be amended to 12%. This will make no difference to the 'All-in Rate' used in the examples.

3 Mechanical Plant

General Note. The pricing of machinery can be contentious, each estimator making different allowances for interest, depreciation, repairs, number of hours the plant works, etc. An example of this could be that for a mixer, although one estimator allows 5 years of life, someone else may allow 10 years; and one contractor may allow for yearly working hours to be 1,750, whereas another contractor may allow only 1,200 hours. How do you allow for wear and tear? These constants when applied can produce a wide variance, and this chapter attempts only to point out the principles involved in pricing machinery.

MACHINERY OWNED BY CONTRACTOR

Working Costs

All calculations in which mechanical plant has to be considered will be based on a figure showing the working cost per hour of the particular piece of equipment.

The factors on which the working costs per hour depend are:

(1) The initial cost of the machine.
(2) The interest on the outlay.
(3) Depreciation.
(4) The cost of repairs and renewals.
(5) The number of hours the plant works per annum.
(6) The running costs: fuel, oil, grease etc.
(7) The operator's wages.
(8) The insurance of the plant.
(9) The licence act, if any.
(10) Transport to site.

(1–3) INTEREST ON THE OUTLAY AND PLANT DEPRECIATION

There are three main methods of dealing with plant depreciation and the interest on capital cost:

(*a*) *Straight Line*. This method presumes that the contractor has a sum of money to invest. It could be invested in stocks and shares to provide an income, or it could be invested in an article of plant to do work and thus make a profit. In this way no interest in outlay is allowed for, as this is considered as being covered in the profit.

(*b*) *Interest on outlay and reserve fund*. This heading speaks for itself, the reserve going towards the replacement of the plant.

(*c*) *Depreciation on the written-down value*. The depreciation is worked out by taking the percentage depreciation on the written-down value of the preceding year. Tax allowances are based on this method.

Examples Assume machinery cost £12,000·00 with a life expectancy of 4 years.

A	Cost of machine	£12,000·00	
	Ddt. Scrap value	£ 400·00	say
		£11,600·00	

Cost per annum = £11,600 ÷ 4 = £2,900·00

Hours worked 1,500 ∴ cost per hour
= £2,900·00 ÷ 1,500 = £1·94

B The example is worked from first principles to explain the method before working out an hourly cost using valuation tables.

Rate of interest, say 10% on both capital and reserve fund.

Interest on Capital Outlay			*Interest on Reserve Fund*
First year	10% on £12,000·00 =	£1,200·00	
Second year	10% on £12,000·00 =	£1,200·00	10% on £3,000·00 = £ 300·00
Third year	10% on £12,000·00 =	£1,200·00	10% on £6,000·00 = £ 600·00
Fourth year	10% on £12,000·00 =	£1,200·00	10% on £9,000·00 = £ 900·00
		£4,800·00	£1,800·00
Ddt. credit interest on reserve fund		£1,800·00	
		£3,000·00	

(i) Average interest costs per annum = £3,000·00 ÷ 4
= £750

$$\therefore \text{ Annual cost per annum} = \frac{£12{,}000{\cdot}00}{4} + 750$$

$$= \underline{\underline{£3{,}750{\cdot}00}}$$

(ii) Using valuation tables,
Cost of machine £12,000·00
Years purchase at 10% single rate for 4 years = 3·170

$$\therefore \text{ Annual cost per annum} = \frac{£12{,}000{\cdot}00}{3{\cdot}170} = £3{,}785{\cdot}00$$

say

N.B. Difference of £35·00 probably due to 3·170 being taken only to third point (i.e. £3,750·00 x 3·170 = £11,888·00)

Cost per hour = £3,785·00 ÷ 1,500 = £2·52.

C The owner can take up to 100% depreciation in the first year depending upon his trading position and this is called the 1st year Allowance. Thereafter he is allowed 25% per annum on the balance, reducing each year on the amount carried forward. This is known as the 'Pool'.

In this example it is taken that it has been decided to claim 60% depreciation in the first year.

		Depreciation
Capital Cost	£12,000·00	
Allow 60% depreciation	£ 7,200·00	£7,200·00
Value of plant after 1st year	£ 4,800·00	
Allow 25% depreciation	£ 1,200·00	£1,200·00
Value of plant after 2nd year	£ 3,600·00	
Allow 25% depreciation	£ 900·00	£ 900·00
Value of plant after 3rd year	£ 2,700·00	
Allow 25% depreciation	£ 675·00	£ 675·00
Value of plant after 4th year	£ 2,025·00	£9,975·00

The average depreciation can be allowed as £9,975·00 ÷ £2,493·75 and treated as before, i.e. cost per hour = £2,493·75 ÷ 1,500 = £1·66.

The difference between **A** and **C** of 28p is accounted for by the asset of £2,025·00 which remains on the books. This may be written off after allowance for scrap value or sold to another firm for that amount. If the life of the plant is realistically assessed at 4 years then the depreciation allowance should be 50% in place of 25% (allowed by tax authorities), and this would reduce the asset to £600 which is close to scrap value.

The above three methods cannot strictly be compared one with the other as

A allows for return on investment to be expressed under Profit and Oncosts,

B works on a replacement basis, and

C is the method which an accountant (e.g. for tax purposes) would probably use and the hourly rate would change from year to year.

For the purpose of this publication the Straight Line method will be used.

(4) COST OF REPAIRS AND RENEWALS

The percentage addition to the cost per working hour varies and only by detailed costing and record-keeping can an accurate figure be obtained.

For estimating purposes the following table may be used:

	% of Cost	*Life of Plant*
Concrete Mixers	$7\frac{1}{2}$ to 10%	6 to 7 years
Dumpers	10 to 15%	3 to 4 years
Bucket-type excavators, etc.	10%	5 to 7 years
Lorries	10 to 15%	3 to 5 years

(5) WORKING HOURS PER ANNUM

Plant is not in constant use: repairs, periods between contracts, weather conditions, etc. all tend to lower the number of working hours. For the purpose of calculating the working cost per hour, therefore, the number of working hours must be assessed. This can really only be arrived at by extracting information from office records.

A lorry may work all year round, but a tractor or scraper may only work 8 months in the year. As the life of plant to a certain extent depends on the number of hours it works, a fair average can usually be arrived at.

A fair number of working hours per annum and the figure which will be used in the examples is 1,500. This includes for standing time.

(6) RUNNING COSTS

Wire ropes, oils and lubricants have to be allowed for (although *normally* included in hire rates for machines). This can be done as a percentage of fuel costs, normally 20 to 25%, or in detail, e.g. lubricating oil for a 5/3½ mixer is 0·045 litres.

Cost of Fuel: Cost of petrol will be taken as 18p per litre. This is allowing for the contractor buying in bulk.

Auto-diesel (for use on roads) at 15p per litre and gas oil (use not allowed on roads) at 9p per litre.

Plant is at present referred to by the trade under 'imperial' names as shown below.

		m^3 *output per hour*	*Litres per hour*	
			Diesel	*Petrol*
Concrete mixer	5/3½	1·00	0·9	1·14
	7/5	1·40	1·5	2·27
	10/7	2·00	1·8	3·41
	14/10	2·80	2·27	4·55
Dumpers	2 c. yd (1·53 m^3)			
	2½ c. yd (1·91 m^3)		2·27	5·68
	3 c. yd (2·29 m^3)			
Mechanical	¼ c. yd (0·19 m^3)		2·27	–
Excavators	⅜ c. yd (0·29 m^3)		3·02	–
	½ c. yd (0·38 m^3)		3·41	–
	¾ c. yd (0·57 m^3)		4·55	–
Lorries	4 ton		2·27	4·00
	6 ton		3·25	5·60

(7) PLANT OPERATOR'S WAGES

Plant operator's wages are part of the working cost of the plant per hour. It is normal to allow for $\frac{1}{2}$ an hour per day for oiling, greasing, etc., and this will be shown separately in the hourly rate build-up (although W.R. allows up to 1 hour per day).

(8) PLANT INSURANCE

Most plant, except for lorries, cars, etc., has small insurance costs as it is covered only for fire and theft. Lorries etc. would have to have third-party risk etc. The insurance is by weight, a 4-ton lorry costing £42·00, a 7-ton lorry £66·00 for third-party cover; or £190·00 for comprehensive insurance on a 7-ton lorry.

$$\text{Insurance cost per working hour of plant} = \frac{\text{Annual premium}}{\text{Working hours of plant}}$$

Say insurance of lorry is £66·00 per annum

$$\therefore \text{Cost per hour} = \frac{£66{\cdot}00}{1{,}500} = 4\tfrac{1}{2}\text{p}$$

(9) LICENCE COSTS

Where plant has to be licensed annually, a charge is made. This can range from a few to hundreds of pounds.

The licence cost of lorries varies for each additional 5 cwt (254 kg). A 4-ton lorry costs £179·95, with a 6-ton lorry costing £323·95 and a $7\frac{1}{4}$-ton lorry costing £413·95.

Say licence of lorry is £197·95 per annum

$$\therefore \text{Cost per hour} = \frac{£197{\cdot}95}{1{,}500} = 13\text{p}$$

The rate of duty is based on the weight of the complete vehicle with no load, and minus such things as tools, spare tyre, fuel and water.

Note: Weights of vehicles may be quoted in this book by either the *ton* or the *tonne*, as the weight difference is nominal and we do not envisage production changes by manufacturers.

MACHINERY HIRED BY CONTRACTOR

The rate of hire can be established by enquiry of plant hiring company or from standard schedules of rates (e.g. those of the Scottish Plant Owners Association) prepared for guidance in hiring out of plant.

The conditions of hire are important and should be examined so that the hirer knows exactly what he is paying for.

The following are some points normally constant with all hiring firms:

(1) The hirer will be responsible for the safe keeping of the plant. Thus the customer is to insure the plant, unless included in an 'all-in' rate. It is normal for the builder to arrange a blanket policy for working on the site to include cover for hired plant.

(2) Where an operator is provided with the plant (which is normal practice, apart from compressors, mixers, etc.), he is deemed to be a servant of the hirer, who is responsible for his actions as if he were in the hirer's direct employ. For example, damage to hidden services is excluded from the responsibilities of the owners. It is assumed that the builder has access to information that will enable him to discover the exact location etc. of drains, cables and various mains.

The only way the hirer could shift responsibility would be to prove that the operator had disregarded clear instructions when causing any damage.

(3) In the event of a breakdown in the plant the owner is entitled to supply, in substitution, plant of a similar type and condition. If the owner is not in a position to do so, the hire is normally terminated and the hirer has *no claims* against the owner for loss or damage sustained by him through this cause.

Puncture or tyre damage suffered on site through nails, glass, etc. is the hirer's responsibility. The hirer is deemed to have knowledge of all such hazards. If it can be shown that an air leak has occurred *solely* through worn tyres, then the owner may be liable.

(4) *Termination of Hire*. The date of commencement is the

date the plant leaves the depot. The date of termination is the date it is returned to the depot or equivalent.

In some circumstances the plant is not returned to the depot if it is convenient to transfer it to another site close by, thus avoiding substantial transporting time. If the hire is not for a fixed period, four days' notice in writing is required from either party for termination.

(5) *Travelling Time*. Time spent in transporting the plant not exceeding one day each way is not normally charged, unless hired for less than one week. If more than one day is so occupied, the remaining time is charged at two-thirds rate. Plant travelling under its own power is charged at full rates irrespective of the period of hire. Transport is the responsibility of the hirer, which means providing the transport. Some owners offer a delivery and collection service, as well as a hire service. Here the user is providing transport by hiring it from the owner.

Example 1

(*a*) A $\frac{3}{8}$ c. yd (0·29 m^3) excavator, crawler-mounted, equipped with a single shovel, costs £12,000·00 and it is assessed to have a life of 5 years, after which it will be sold for scrap value of approximately £500·00. What is the hourly rate?

Cost of machine	£12,000·00
Deduct Scrap value	£ 500·00
	£11,500·00
Cost per annum = £11,500·00 ÷ 5 =	£ 2,300·00
Hours worked 1,500;	
∴ Cost per hour for interest and depreciation = £2,300·00 ÷ 1,500	153p
Repairs and renewals 10%	15p
Running costs: fuel 3·02 litres @ 9p per litre	27p
oil and grease 20% of 27p	6p
Plant insurance and licence costs: premiums ÷ 1,500, say	9p
	210p

Labour

Under Working Rule 3B8.1 the driver of a $\frac{3}{8}$ c. yd excavator receives a labourer's rate plus £1·00 per week	
∴ Driver @ 190p plus $2\frac{1}{2}$ = $192\frac{1}{2}$p per hour plus Attendant @ 190p per hour	$382\frac{1}{2}$p
Cleaning and maintaining machine $\frac{1}{2}$ hour per 8-hour day:	
∴ $\frac{1}{16}$ of $192\frac{1}{2}$p =	·12p
Cost of excavator and labourer per hour	$604\frac{1}{2}$p

Note: Transport is allowed for later in the unit-rated items.

(*b*) The same plant hired for a week of 40 hours will cost £156·00

Hire of machine	£156·00
Labour	
Driver @ 190p x 40 hours plus £1·00	£ 77·00
Attendant @ 190p x 40 hours	£ 76·00
C/fd	£309·00

Machine will be used for 37 hours, allowing 3 hours 'standing time' per week, i.e. running repairs, inclement weather, averaged at approximately $\frac{1}{2}$ hour per day.

∴ Cost per hour = £309·00 ÷ 37 =	835p
Cleaning and maintaining $\frac{1}{2}$ hour per 8-hour day	
∴ $\frac{1}{16}$ of $192\frac{1}{2}$p	12p
Fuel	27p
Insurance (licence cost included in hire rate)	2p
Oil and grease (included in hire rate)	–
Repairs and renewals, etc. (included in hire rate)	–
	876p

From the above examples (*a*) and (*b*), it can be seen that it is cheaper to own the plant, but this is only true if the turnover of work for the machine justifies a figure of 1,500 hours or similar. Most small builders find it better to hire certain pieces of machinery.

Type of plant required

Plant output required =

$$\frac{\text{Amount of work}}{\text{Plant hours necessary to carry out work}}.$$

E.g. the excavations of a contract are to be carried out in 125 days according to the programme of work, and the volume of excavations to be removed is in the region of 11,000 m³.

$$\text{Plant output required} = \frac{11{,}000}{125 \times 8 \text{ hours in working day}}$$

$$= 11 \text{ m}^3 \text{ per hour.}$$

Bucket capacity of excavator in m³	m³ per hour		
	Dragline excavator below level of tracks	Backacter drain excavation	Skimmer exc. level or above tracks
0·19 ($\frac{1}{4}$ c. yd)	8·40	7·65	9·20
0·29 ($\frac{3}{8}$ c. yd)	13·75	10·70	15·30
0·38 ($\frac{1}{2}$ c. yd)	19·10	16·80	20·60

From the above table a $\frac{3}{8}$ c. yd (0·29 m³) digger would be used as tables do not allow for standing time.

Note: Output as above has nothing to do with cost per hour, e.g., if allowing for standing time, calculation should be, say,

$$\frac{11{,}000}{125 \times 7\frac{1}{2} \text{ hours}} = 11{\cdot}73 \text{ m}^3.$$

All that will appear in the Bill of Quantities will be a description of the type of excavation and a quantity in either m² or m³. It is the estimator's job to decide how the work is to be done.

Having decided what excavator to use, he must find out the number of lorries and/or dumpers required to keep it constantly at work.

$$\text{No. of lorries/dumpers} = \frac{\text{Time to load and haul}}{\text{Time to load}}$$

Example 2

What would be the cost of excavating from a temporary spoil heap with a $\frac{1}{2}$ c. yd (0·38 m^3) mechanical excavator fitted with a skimmer, the material being hauled in 5-ton tipping lorries, to be removed off site to a tip 1 mile away? It is estimated that the excavator will take out approximately 20 m^3 per hour. Two men attend the excavator, employed on trimming the formation, and one man is on the tip. It is assumed that 'remove off site' will include for taking from temporary spoil heaps.

$\frac{1}{2}$ c. yd (0·38 m^3) excavator takes 3 minutes to dig 1 m^3
5-ton lorry will haul, say, 3 m^3 per load
Time to travel 1 mile, unload and return: say lorry travels at 12 miles per hour, time will be

$$\frac{60}{12} \times \frac{2}{1} = 10 \text{ minutes}$$

plus unloading at tip, say 8 minutes

18 minutes

Time to load lorry = 3 x 3 = 9 minutes

$$\therefore \text{Number of lorries} = \frac{9 + 18}{9} = 3 \text{ lorries}$$

Assume the following rates:

Lorries £4·00 per hour, ∴ three lorries	1200p
Excavator £6·50 per hour	650p
3 attendants @ 190p	570p
	2420p

Output of excavator is 1 m^3 in 3 minutes;	
$\therefore$ Cost per m^3 = 2420p × $\frac{3}{60}$ =	121p
Cost of transport to site and removal at completion is, say, £30·00, with quantity of excavation 600 m^3;	
Cost = $\frac{£30{\cdot}00}{600}$ =	5p
Local authority tipping charge, say 45p per tonne @ 2 tonnes per m^2	90p
(in some districts by the load)	
	216p
Profit and Oncost 20%	43p
	259p
$\therefore$ Rate per m^3	£2·59

Further examples of machinery costs will be found under the three chapters on Excavations, Concrete Work, and Brickwork, Chapters 5 to 7.

4 Preliminaries

The following is the Preliminaries Section of a recent Bill of Quantities followed by a clause-by-clause explanation.

Preliminary Particulars

Projects, parties and consultants

The project is an extension to an existing old peoples home known as Lochan Eventide Home located at Oldchurch, Fife.

General Description of the Works

The works comprise a 25 bedroom two storey extension in one wing to an old people's home of traditional construction together with extensive alterations to parts of the existing building and the addition of a lift shaft, etc. all as shown on the accompanying drawings.

Employer

The employer is the Society for Social and Moral Welfare, C/o Smith & Robertson, 14 Rutland Crescent, Glasgow.

Architect

The Architect is Aitken Architects, Richmond Gardens, Dundee.

Quantity Surveyor

The Quantity Surveyor is A. Measurer, Chartered Surveyor, Newtown, Scotland.

Description of Site

The site is as shown on the location plan, indicating its boundaries, means of access and position of the Works, included with these Bills of Quantities, situated approximately 500 m on the south side of the village of Oldchurch. Free road access to the site as indicated on plan.

Tenderers are recommended to examine the drawings and visit the site and fully satisfy themselves regarding local conditions, accessibility of the property, the full extent and character of the operations, the nature of the layout, the supply and conditions affecting labour and materials and the execution of the contract generally. No extras will be allowed on account of any omission or error arising from lack of such knowledge.

C/fd

Drawings

The following drawings which were used in the preparation of the Bills of Quantities will form part of the contract and a copy of each is issued with the tender documents.
Site Plan 10K4/49; Location Drawings 10K4/15 to 21 inclusive.
The following further drawings used in the preparation of the Bills of Quantities can be seen at the Architect's Office by appointment:
Drawings Nos.: 10K4/5, 15D, 22, 25, 30 to 34 inclusive and 37 to 48 inclusive.

Tender

The Employer is not bound to accept the lowest or any tender.
The Contractor on acceptance of his offer shall proceed immediately with the preparation of a programme or statement which shall clearly set forth the sequence of all operations and the time limits within which the Contractor proposes that each operation shall be commenced and completed. The Contractor, in the preparation of this programme, shall be held to have consulted and co-ordinated the whole of the works embraced in this Contract including the work of all sub-contractors, nominated or otherwise. On agreement or negotiated amendment of the programme by the Architect, the Contractor shall be responsible for the execution of the works in conformity therewith.

Overtime

Where overtime is *specifically* ordered in writing by the Architect, the net difference between flat time and overtime rates plus a percentage for profit and overheads (Contractor to state here percentage required) will be allowed (other than that carried out under Daywork) provided detailed returns are submitted each week.

Tendering Procedure

The tender together with the priced Bill of Quantities under separate sealed covers enclosed 'Tender or Bill of Quantities for Additional Accommodation at Lochan Eventide Home, Oldchurch, Fife' to be lodged with the Architect, Richmond Gardens, Dundee.

Apprentices

In the case of any apprentice employed in the execution of this Contract, the Contractor shall comply with the terms of the Apprenticeship Scheme of the Building Industry in Scotland.

C/fd

Method of Measurement

The Bills of Quantities are prepared in accordance with the Standard Method of Measurement of Building Works, Sixth Edition, with the exception of the circuit wiring which has been numbered on a points basis.

Contract

Form, Type and Conditons of Contract

The works embraced in this Contract are to be carried out in accordance with the Schedule of Conditions of Building Contract (Private Edition with Quantities) 1963 Edition (July 1977 Revision) as amended and modified by the provisions contained in the Scottish Supplement, July 1977 and the Abstract of the said Schedule of Conditions (Appendix II) and the provisions for payment of V.A.T. forming Appendix II to the Building Contract, all of which conditions amendments and supplementary clauses are held to be incorporated in and form part of the Contract Bills. The Building Contract shall be deemed to be executed when letters of acceptance have been exchanged between parties and accordingly at Clause 1 of the Scottish Supplement the meanings at 2 and 11 shall be deleted and the following substituted:

2. Articles of Agreement – Tender and letter of acceptance thereof
11. Execution of Contract – Acceptance of Tender

Schedule of Clause Headings

Clause numbers and headings of the Conditions of the Standard Form of Building Contract 1963 edition (July, 1977 Revision) incorporating Clause 1 of Scottish Supplement, July 1977.

1	Contractor's Obligations ______		
2	Architect's Instructions ______		
3	Contract Documents (Clause 3(2)(a) is deleted) ______		
4	Statutory Obligations, Notices, Fees and Charges (items for rates on temporary buildings and water for the works are provided elsewhere) ______		
5	Levels and setting out the works ______		
6	Materials, goods and workmanship to conform to description, testing and inspection ______		
7	Royalties and Patent Rights ______		
8	Foreman in charge ______		
9	Access for Architect to the Works ______		
10	Clerk of Works ______		
11	Variations, provisional and prime cost sums (see Appendix II) ______		
12	Contract Bills ______		
13	Contract Sum ______		
13A	Value Added Tax (*see* Scottish Supplement) ______		
14	Unfixed Goods and Materials (as amended by the Scottish Supplement Clause II) ______		
15	Practical completion and defects liability (*see* Appendix II) ______		
16	Partial possession by Employer (do.) ______		
		C/fd	

	B/fd			
17	Assignment or Sub-Letting (The Contractor will be held to employ those sub-contractors stated in the list attached to the back of this document)			
18	Injury to persons and property and Employer's Indemnity			
19	Insurance against injury to persons and property (*see* Appendix II)			
	Clause 19(2)(a) shall not apply.			
19A	Excepted risks, nuclear perils etc.			
20	Insurance of the Works against Fire, etc. (*see* Scottish Supplement) Clause (20)(c) of the Schedule of Conditions shall apply			
21	Possession, completion and postponement (*see* Appendix II)			
22	Damages for non-completion (do.)			
23	Extension of time. Sub-Clause (J) is to apply			
24	Loss and expense caused by disturbance of regular progress of the work			
25	Determination by Employer (as amended by Scottish Supplement Clause V)			
26	Determination by Contractor (as amended by Scottish Supplement Clause VI) (*see* Appendix II)			
27	Nominated Sub-Contractors (as amended by Scottish Supplement Clause VII) (*see* Appendix II)			
28	Nominated Suppliers (as amended by Scottish Supplement Clause VIII)			
29	Artists and Tradesmen			
30	Certificates and Payments (as amended by Scottish Supplement Clause IX) (*see* Appendix II)			
31	Fluctuations Clauses 31A, 31C, 31D and 31E shall apply. The Contractor is to enter in the list attached to each Bill the market prices of materials and goods to which Clause 31A(c) is to apply and the application thereof shall be restricted to the materials and goods so listed			
32	Outbreak of hostilities			
33	War damage			
34	Antiquities			
35	Arbitration (as amended by Scottish Supplement Clause XI)			
	Clause XII of Scottish Supplement. Contracts let on a separate trades basis. This clause shall be deleted			
	Limitations of Working and Order of Work			
36	The Contractor will be required to keep within the boundaries as indicated on the site plan.			
	1. Lay main house drains clear of new extension and connect to existing drains at both ends.			
	2. Demolish out-buildings in courtyard area and proceed with the work on the extension on two floor levels, including the construction of the Lift Shaft and the re-construction of the work in the			
		C/fd		

	B/fd			
	courtyard area. At the same time the work on the ground floor S.E. bedroom, mortuary and fuel store to be carried out. 3. Occupy new extension and proceed with work on first floor, existing hospital wing and existing Superintendent's house. 4. Carry out minor alterations to existing building, i.e. kitchen, new exit door from existing dining room, new fire exit door on ground floor at N. end of building. Install the lift and break-through at each floor at a time convenient to the working of the Home. In the carrying out of any section of work it is found necessary and authorised by the Architect to work outside the normal working hours, it will be paid for as defined under the 'overtime clause'.			
37	Allow for protecting, upholding and maintaining common pipes, ducts, sewers, service mains, overhead cables, etc. during the execution of the works. The Contractor is to make good any damage due to any cause within his control at his own expense or pay any costs and charges in connection therewith ———			

General Facilities and Obligations

	Pricing		
38	Allow for all necessary plant, tools and vehicles for the work		
39	Do. for complying with the Building (Safety Health and Welfare) Regulations in respect of all workpeople employed (including those employed by Nominated Sub-Contractors) on the site		
40	Do. for disbursements arising from the employment of workpeople		
41	Do. for transport for workpeople		
42	Do. for site, administration, and security		
43	Do. for maintenance of public and private roads		
44	Do. for control of noise, pollution and all other statutory obligations		
45	Allow for providing water for the works and temporary arrangements for storing and distributing about the site		
46	Do. for providing lighting and power for the works and temporary arrangements for distributing about the site and for lighting to hoardings and the like		
47	Temporary roads, tracks, hardstandings, crossings and the like including use by Nominated Sub-Contractors		
48	Temporary sheds, offices, messrooms, sanitary accommodation and other temporary buildings for use of the Contractor. Contractor to provide area for Clerk of Works (within own offices) for displaying plans		
49	Rates on temporary buildings		
50	Temporary telephone facilities on the site for the use of the Contractor and Nominated Sub-Contractors and paying all accounts for do. (Calls made by Clerk of Works will be paid for separately)		
51	Temporary hoardings and gantries. Temporary fencing, hoardings, fences, planked footways, guard rails, gantries and the like for the proper execution of the work, for the protection of the Public and Occupants of the adjoining premises and for meeting the requirements of any local or other authority		
52	Temporary scaffolding for the proper execution and completion of the works		
53	Protecting the works from inclement weather		
54	The Contractor is to provide all necessary heating and attendance for drying out the buildings and controlling humidity immediately before handing over and at such other times as may be necessary to facilitate the progress and completion of the works including the work of all sub-contractors, and is to deliver up all parts of the buildings in a thoroughly sound, dry and perfect condition and to ensure that any damaged items have been replaced. The Contractor will be permitted to use the new heating system for this purpose, *if available*, providing he indemnifies the sub-contractor for Heating Services, gives all notices, pays all charges, including attendance and for adapting the system as necessary		
		C/fd	

		B/fd	
55	Provide for removing all rubbish and debris from the site and cleaning the works internally and externally. On completion, the works shall be cleaned which shall be deemed to include scrubbing floors, washing pavings, polishing glass inside and outside; cleaning sanitary fittings; flushing drains and manholes; cleaning gutters and down pipes; leaving the whole of the new premises and areas of alteration in the existing buildings clean and ready for occupation ______		

APPENDIX NO. II: ABSTRACT OF SCHEDULE OF CONDITIONS

(Appendix I in the Scottish Supplement re-defines various words and phrases mainly to align with Scottish Law. Appendix II is otherwise identical to the Appendix in the Standard Conditions of Contract.)

Defects Liability Period

Clauses 15, 16 and 30 12 months insurance cover for any one occurrence or series of occurrences arising out of one event. Clause 19(1) (a) £500,000

Date for Possession

Clause 21 within 14 days of acceptance of tender

Date for Completion

Clause 21 12 months from date for possession

Liquidated and Ascertained Damages

Clause 22 At the rate of £200 per week or part thereof

Period of Delay

(i) by reason of loss or damage caused by any one of the contingencies referred to in Clause 20(A) or 20(B) NOT APPLICABLE

(ii) for any other reason 1 month

Prime Cost Sums for which the Contractor desires to Tender

Clause 27(g)

Any tender submitted by the Contractor shall, for the purposes of discount and profit, be treated as if it were submitted by a third party.

Period of Interim Certificate

Clause 30(1) 1 month

Retention Percentage 5%

Period of Final Measurement and Valuation

6 months from Date of Practical Completion

Percentage Additions

Clause 31(E) 20%

Amount of Preliminaries carried to Summary £

Description	%	£	p
Prime Cost, Provisional Sums, and Daywork, etc.			
Allow Prime Cost Sum of £10,000 for Heating Installation carried out by Nominated Sub-Contractor		10,000	00
Contractor's Profit on do.			
Allow for General Attendance on do.			
Allow Prime Cost Sum of £9,500 for Electrical Installation carried out by Nominated Sub-Contractor		9,500	00
Contractor's Profit on do.			
Allow for General Attendance on do.			
Allow for Prime Cost Sum of £5,000 for Painter Work of Existing Building carried out by Nominated Sub-Contractor		5,000	00
Contractor's Profit on do.			
Allow for General Attendance on do.			
Allow Prime Cost Sum of £7,500 for Lift Installation carried out by Nominated Sub-Contractor		7,500	00
Contractor's Profit on do.			
Allow for General Attendance on do.			
Allow Prime Cost Sum of £6,000 for Septic Tank Installation carried out by Nominated Sub-Contractor		6,000	00
Contractor's Profit on do.			
Allow for General Attendance on do.			
Allow Prime Cost Sum of £2,000 for Precast Concrete Floor Units carried out by a Nominated Sub-Contractor		2,000	00
Contractor's Profit on do.			
Allow for General Attendance on do.			
Provisional Sums			
Provide Provisional Sum of £25 for telephone calls made on behalf of the Employer		25	00
Dayworks			
Allow the sum of £3,500 for the amount of wages in daywork as defined in Clause 11(4)(c)(i)		3,500	00
Allow for profits and overheads			
Allow the sum of £700 for the cost of materials in daywork as defined in Clause 11(4)(c)(i)		700	00
Allow for Profit and overheads			
Allow the sum of £400 for the use of mechanically operated plant and transport for the time engaged in dayworks		400	00
Allow for Profit and overheads			
Clause 11(4)(c)(ii)			
*(1) Labour	%		
(2) Materials	%		
(3) Plant	%		
*(1) Labour	%		
(2) Materials	%		
(3) Plant	%		
*State specialist trade applicable.			
Prime Costs, Provisional Sums, and Dayworks carried to Summary	£		

The pricing of the Preliminaries must be carried out with care. The estimator must have a complete understanding of the conditions of contract to allow him to price the document properly.

The Preliminaries Section is divided into eight sub-sections.

(1) Preliminary Particulars

This section is to inform the contractor on some basic points. The name of employer architect, quantity surveyor and others who may be of importance to the contractor. It informs him as to where the work will be carried out, the type of construction involved and various limitations that will be imposed. In other words it attempts to set out a general picture of the work on which the contractor will submit his tender.

This information is usually typed on ruled paper to allow prices to be inserted, if the contractor so wishes, against each clause. These clauses are not normally priced. Overtime is accounted for in the 'All-in' rate.

(2) Contract

Particulars of the form and type of contract must be given (B4). If the conditions of contract are standard it is sufficient to give a schedule of clause headings giving particulars of the edition to be used. For other conditions of contract the conditions are to be set out in full or a schedule of clause headings is to be given where a copy of the full conditions are to be supplied with the bill.

Item Nos.

1 to 4 incl. Not normally priced.

5 Usually allowed for in the overheads but could be priced as below.

E.g. building size 48 x 19 m on plan.

Material	say 38 x 38 mm pegs 450 mm long 80 @ 10p =	800p
	profiles 20 @ £3·00 =	6,000p
	nails, setting out lines, etc. say £1·00 =	100p
	C/fd	6,900p

Item Nos.			
		B/fd	6,900p
	Labour 30 hours of labourer @ 190p =		5,700p
			12,600p
		Profit and Oncost 20%	2,520p
			15,120p
		Rate would be	£151·20

6 and 7 Not priced.

8 Although this is a requirement regarding the presence of a competent person, if the item is priced, other supervisory staff wholly employed on the contract can be included. Thus, a site agent plus one or two others could easily cost the contractor £250·00 per week. Thus for a contract period of 100 weeks this item could cost £250·00 x 100 plus 20% for Profit and Oncost = £30,000·00. Surveyors, visiting supervisors, engineers, etc. are allowed for in overheads, but this must largely depend on the nature of the job and the extent to which staff are wholly committed.
This is an important item if the contract period is varied as it is time related and thus subject to pro-rata adjustment.

9 to 18 inclusive Not priced. Item 18 has no requirement to be priced.

19 Part of Clause 19(1)(a) can be covered in the 'All-in' rate and that to do with property could be the premium required for that particular contract divided by the number of weeks in the contract period. The contractor may have a comprehensive policy to cover all his work and in this case the amount involved is likely to be allowed for in overheads.
Clause 19(2) is for special requirements covered by

Item Nos.

a Provisional Sum to be entered by the client in the Bill of Quantities.

20 This clause covers insurance against fire, etc. and it is only in the case of alternative 20(a) that this item is priced by the contractor.
This clause should be read in conjunction with the Appendix to the contract to ascertain the percentage to be added to cover professional fees. The figure to set against this item is usually derived from an insurance company's quotation. An average premium is 20p per £100·00 p.a.
As this item is value-related, premiums vary according to contract value, it may be adjustable in the event of variations occurring.

21 to 30 inclusive Not priced. Item 22 Damages for non-completion should be read with the Appendix to determine the amount of damages payable. Although not priceable here, this item would have to be considered by the contractor in determining the level of overtime working required to meet his contract commitment – allowed for in the 'all-in' rate.

31 The fluctuations clause almost requires a book to itself and we shall not enter into any explanation other than to say that generally speaking Clause 31A covers fluctuations in labour rates, in governmental impositions or the cost of any new imposition payable by the contractor in his capacity as an employer, and in the prices of specified materials, and Clause 31B covers fluctuations in governmental impositions as in part A and in the prices of specified materials due to changes in tax or duty.
Parts A and B are *alternatives* and it is bad practice to attempt to omit Clause 31 from the contract. Clause 31(F) is used where the parties have agreed that fluctuations should be dealt with by adjustment using the National Economic Development Office Price Adjustment Formula.

Item Nos.

Clause 31(B) together with (C), (D) and (E) is the nearest one can get to a firm price offer.
Increases can be allowed for in the 'All-in' rate if 31(B) is in operation as follows:

1. Date of Tender as defined under Clause 31) — 11 October, 19–.
2. Starting date of contract — 20 November, 19–.
3. Contract period — 40 weeks (24 August 19–.)
4. Known increases
   6.11.19–. Craftsmen 2½p
   Labourers 1½p per hour
5. Estimated increases
   5.2.19–. Craftsmen 2p Labourers 1½p
   6.5.19–. Craftsmen 2½p Labourers 2½p

	C	L
Period 20.11.19–. to 4.2.19–.		
11 weeks × 2½p / 1½p	27½p	16½p
Period 5.2.19–. to 5.5.19–.		
13 weeks × 2 p / 1½p	26 p	19½p
Period 6.5.19–. to 24.8.19–.		
16 weeks × 2½p	40 p	40 p
	93½p	76 p
40 week period ∴ average hourly increase	234p	190p

∴ Increase cost per week is 93½p and 76p (40 hour week)
∴ Weekly rate for Tradesman and Labourer would be based on £52·93½ and £45·96 in place of £52·00 and £45·20 used in the 'All-in' rate.

32 to 37 inclusive. Not priced.

Item Nos.

(3) General Facilities and Obligations

Rule B13 of the S.M.M. lists items which for convenience in pricing must be given

38 (1) *Plant, tools and vehicles*

Certain types of plant are best related to the unit rates in the bill of quantities, e.g. excavators and mixers, dumpers, etc., but other plant is in such general use that it is better allowed for here.

Example A

Hoist: 2 required one for 15 weeks and one for 17 weeks.

Weekly hire rate of hoist £21·60 =	2,160p
Fuel, oil, etc.	180p
Labour £76·00 plus 60p =	7,660p
Cleaning and maintaining ($\frac{1}{2}$ hour per day) $\frac{1}{16}$ of $\frac{£76{\cdot}60}{40}$	479p
	10,479p
Profit and Oncost 20%	2,096p
1 Week costs	12,575p

∴ 2 machines for 16 weeks (average) = 12,575p x 2 x 16 £4024·00

Example B

Tower Crane. Metre-Tons capacity 50

Weekly hire rate £253·00	25,300p
Fuel, etc. included.	
Labour £76·00 plus 160p additional plusage if over 50 ft from ground.	7,760p
Slinger £76·00 plus 60p	7,660p
Cleaning and maintaining ($\frac{1}{2}$ hour per day) $\frac{1}{16}$ of £77·60	485p
1 Week	41,205p

Item Nos.

	∴ 16 weeks costs £412 say x 16 =	£ 6,592·00
	Prepare ground and lay 15 m of track say	£ 600·00
	Rail track hired @ 55p per m per week (55 x 15 x 16)	£ 132·00
	Transport to and from site	£ 450·00
	Erection and dismantling	£ 540·00
	Electrical supply installation	£ 250·00
		£ 8,564·00
	Profit and Oncost 20%	£ 1,713·00
		£10,277·00

When calculating the cost of small plant and tools some builders allow a small percentage.
E.g. 2½% for one category of work and 1% for another.

39 *Safety, Health and Welfare of Workpeople*—see Temporary Sheds, etc.

40 and 41 *Covered under 'All-in' Rate.*

42 *Site administration and security*

This will depend upon (1) location of site and (2) type and size of contract. If, for example, a watchman was considered necessary the following would have to be allowed for.

(1) Total cost of watchman x number of weeks required.
(2) Cost of watchman's hut, fuel, brazier, etc.
(3) Cost of warning lights and barriers, etc.

Alternatively a security firm may be employed and the charges made for this service allowed against the item.

Note that the above is at the Contractor's risk. If watching is specifically required by the employer it must be stated under S.M.M. Clause B(8).

Item Nos.

43 *Maintenance of Public and Private Roads*
The estimator must assess the possible damage caused by his vehicles and should at least allow for cleaning roads and pavements and keeping the place tidy.

44 Not priced.

45 *Water for the works*
There are three components of this price:
(*a*) Water Board charges for building purposes – In the writer's area the charges are:
(i) 25p per £100 of the total contract value for traditional building and
(ii) 12½p per £100 of the total contract value for non-traditional building.
These charges to be paid in advance on making applications for a building supply and to be based on all dry and wet trades, prefabricated work and pre-mixed concrete etc. The charges to be applied to all cases in general, but no charge shall be applied where the total value of the work is less than £1,000.
(*b*) Cost of provision of standpipe and hose by plumber.
(*c*) Cost of temporary receptacles.

46 *Lighting and Power for the Works*
The contractor must allow for the following:

(*a*) Electricity authority charges.
(*b*) Cost of wiring and lighting and power points about the site.
(*c*) Cost of lamps, etc.
(*d*) Cost of transformer to provide safe working voltages.
(*e*) Dismantling and removal.
If no supply is available, either the cost of generators or of Calor gas equipment must be allowed.

Item Nos.

47 *Temporary roads, etc.*

Each job must be examined and suitable hard-standings etc. allowed for. Where appropriate it is normal to put in the metalling for a road at the start of the job and make good near the end of the contract before putting down the finishing layers. If this can be achieved there will be little allowed under this item.

48 *Temporary sheds, etc.*

Certain minimum standards are laid down in various government legislation and come under Clause B.13.1.S of the S.M.M.

Example

Two offices required for a contract period of 52 weeks.

Initial cost of 2 offices £910·00		
Life span of offices estimated at 7 years,		
∴ Cost per annum		£ 130·00
Haulage to and from site	say	£ 45·00
		£ 175·00
Labour		
Erect and dismantle 80 hours @ 190p		£ 152·00
Lining floor, walls, and ceiling		£ 150·00
Decorations (touching up)		£ 25·00
Foundations and sleeper walls		£ 120·00
Electrical Installation		£ 195·00
Heaters 4 @ £28·00		£ 112·00
Power supply £10·00 per week, ∴ 52 x £10·00		£ 520·00
Furniture		£ 56·00
Cost of 2 offices		£1,505·00

Messrooms, stores, and sheds are calculated on a similar basis.

Toilets vary according to site conditions. Sometimes a chemical closet is used including allowance

Item Nos.

for regular servicing and sundries (soap, towels, disinfectant). Connection to the main drainage would be priced on the basis of approximate quantities.

49 *Rates on Temporary Buildings*

Local Authorities can levy rates on site buildings if they are in a position over a period of time. An indication of rateable value should be obtained and an allowance made. Say 35p per m^2 per annum

$\therefore$ Cost would be Area x Rate x $\frac{\text{Rate in £}}{\text{£}}$ x Period in years:

say $100\ m^2 \times 35p \times \frac{140p}{100p} \times 1$ year (contract of 52 weeks) = £49·00

The 140p allowed could radically differ from region to region.

50 *Telephone Facilities*

If not exceeding 1 year, these will be charged on a temporary line basis. If over 1 year, they will be charged on normal installation and rental basis. In both cases calls are charged at normal private line rate.

The cost of calls made on behalf of the employer is a separate item and is the subject of a Provisional Sum.

51 *Temporary Screens, Hoardings and the like*

Here it is necessary to prepare approximate quantities to be priced up in the normal manner, allowance being made for subsequent dismantling and for 'use and waste' of materials. There may be a local authority 'Hoarding Permit' charge based upon the area of footway occupied (in the writers' area approximately 5p per m^2 per week). Alternatively, 'free' hoardings may be negotiated with an advertising firm, in which they will erect

Item Nos.

hoardings free of charge in exchange for advertising rights. It is to be noted, however, that the permission of the employer should be first obtained as the sole right of advertising is normally reserved to the employer. Also, local authority (planning) permission for this type of hoarding may have to be obtained.
Note also to allow for appropriate safety lighting, guard rails, warning notices, and the like.

52 *Scaffolding*

The contractor is able to assess the scaffolding required from the preliminary particulars. In certain cases additional information must be given.

Example

The term 'sheeting square' was used in the trade to refer to an elevational area of 10′0″ x 10′0″ which will be taken as 3 x 3 m (approx. 97 sq. ft) for the purposes of these examples.

General

The average length of scaffolding in a sheeting square of normal access scaffolding with standards at 2·44 m centres, every lift handrailed, is 38 m. Bricklayers' scaffolding with standards at 1·82 m centres has 45 m per sheeting square.

Output

Rate of erection, 442 m per day for a squad of 3 men. Rate of dismantling, 884 m per day for a squad of 3 men.

Weights		
	Steel tube	216 m per tonne
	Alloy tube	642 m per tonne
	Fittings (mixed)	985 per tonne
	Boards	59 per tonne

It is intended to erect a standard scaffold (as shown in the diagram on page 70) 24·40 m long x 11·00 high, having 1 lift boarded, for 20 weeks.

Erected by Contractor using own Labour and Material

Material

Standards	11 @ 12·00 m	132·00 m
	11 @ 11·00 m	121·00 m
Ledgers	2 x 6 lifts x 24·40 m	292·80 m
Transomes	11 x 6 lifts x 1·52 m	100·32 m
Boardbearers	15 x 6 lifts x 1·52 m	136·80 m
Handrails	6 lifts x 24·40 m	146·40 m
Stop-ends	2 x 6 lifts x 1·52 m	18·24 m
Internal bracing	6 x 6 lifts x 2·74 m	98·64 m
External bracing	3 x 2 x 6·10 m	36·60 m
Ties	8 x 2·74 m	21·92 m
		1,104·72 m

Steel tube 1,104·72 m @ 110p per m =	121,519p	
Fittings 640 @ 102p each	65,280p	
Boards (including toe board)	15,750p	
30 @ 525p each	15,750p	
	202,549p	
Waste $2\frac{1}{2}\%$	5,064p	
	207,613p	
Number of uses say 30, ∴ Cost per use = 207,613p ÷ 30 =		6,920p

Labour

Erecting: 3 men take 2 days x 8 hours	48 hours	
Dismantling: 3 men take 1 day x 8 hours	24 hours	
	72 hours	
72 hours @ 220p =		15,840p
	C/fd	22,760p

	B/fd	22,760p
Transport		
Tube 216 m per tonne,		
∴ 1,104·72 m weigh	5·12 tonnes	
Fittings 985 per tonne,		
∴ 640 weigh	0·65 tonnes	
Boards 59 per tonne,		
∴ 30 weigh	0·51 tonnes	
	6·28 tonnes	
6·28 tonnes both ways = 12·56 tonnes @ 220p (from records)		2,763p
Maintenance		
2 hours per week for 15 weeks @ 220p		6,600p
(No maintenance first 4 weeks and last week)		
		32,123p
Profit and Oncost 20%		6,425p
		38,548p
Scaffolding for 20 weeks, say		£386·00

Erected by Specialist Contractor

4 weeks minimum length of hire and rates quoted for this with an additional amount for every week thereafter.

Material

Steel tube	1,104·72 @ 3p per m =	3314p
Fittings	640 @ $2\frac{1}{2}$p each =	1600p
Boards	30 @ 54p each =	1620p
		6534p

On contract 50% or less of full hire would be charged	3,267p
Labour	
Erect and dismantle 72 hours @ 'all-in' rate of £3·20	23,040p
Expenses	
Assume 40p travelling expenses per man per day 3 x 3 x 40p	360p
Transport	
12·56 tonnes of material @ £2·20 per tonne	2,763p
	29,430p

This price of £295·00 would be for the hire of the scaffold for 4 weeks.
If a sheeting square is taken as 3·00 x 3·00 m = 9m^2 then the number of squares in 24·40 x 11·00 m =

$$\frac{24{\cdot}40 \times 11{\cdot}00}{9{\cdot}00} = 29{\cdot}82$$

Price per sheeting square would be £295·00 ÷ 29·82 = £9·89.
The above price is for 1 lift boarded; each additional lift would cost:

Boards 1,620p @ 50% hire rate	810p
Labour 2 hours @ 320p per hour	640p
Transport 1 tonne say @ £2·20	220p
	1,670p
	or 56p per square

Thus cost of

(*a*) 1 lift boarded £295·00 = £9·89 per square
(*b*) 2 lifts boarded £311·70 = £10·45 per square
(*c*) 3 lifts boarded £328·40 = £11:01 per square

If the period was extended beyond 4 weeks the additional cost per square per week would be

£32·67 ÷ 4 weeks = £8·17
£8·17 ÷ 29·82 = 27p

(*a*) 1 list boarded 27p per square per week on £9·89
(*b*) 2 lifts boarded 34p per square per week on £10·45
(*c*) 3 lifts boarded 41p per square per week on £11·01

Thus to hire scaffolding for 20 weeks would cost £295·00 plus 16 times 27p x 29·82 = £295·00 + £128·82 = £423·82.
According to the trade the additional costs include for maintenance.
The comparison of the hired cost of £423·82 with the main contractor's own cost of £386·00 can only be made if the main contractor has a sufficient turnover of scaffolding work to warrant buying material and employing scaffolders full-time. Even so, on short-term work the hiring contractor can still seriously challenge his rates. Thus many contractors have decided that there is very little, if any, advantage in owning their own scaffolding and consequently there are now a number of firms which deal almost purely in the hire of scaffolding.
The above rates could be applied to a height of 21·00 m. Thereafter an increase of approximately 20% for every 3·65 m increment in height would apply.
If the contractor decided to utilize his own labour to erect and dismantle, the preferred rates as shown in the example would not apply.
Thus the rate per square for straight hire would be:

(*a*) Average scaffold 1 lift boarded 54p per square per week.
(*b*) Average scaffold 2 lifts boarded 68p per square per week.
(*c*) Average scaffold 3 lifts boarded 82p per square per week.

These prices may be converted to a square metre rate if desired.

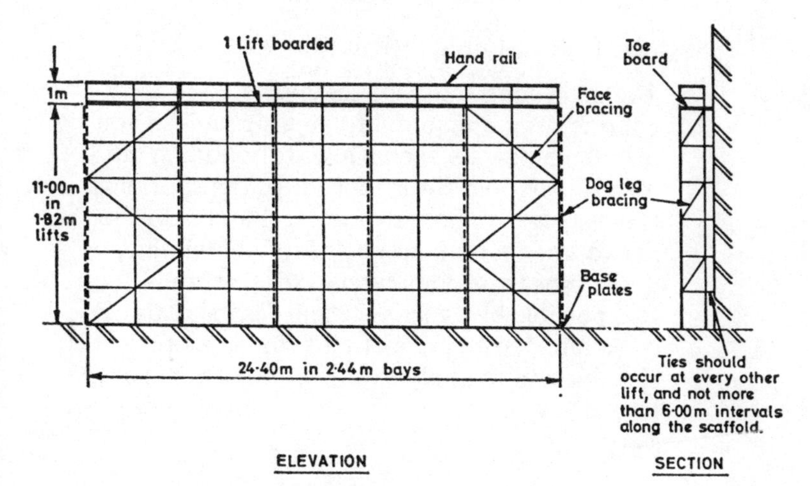

Item Nos. (5) Protection, Drying, and Cleaning

53 to 55 *Protection*

This only covers protection from inclement weather, protection from damage by other trades being given in the trade bills where appropriate. The extent of protection to be allowed for varies but this is usually nominal (e.g. sacking to top of new brickwork and straw covering to concrete during frosty weather).

N.B. Special protection (temporary roof, etc.) where specifically required is given under S.M.M. Section C and is priced on approximate quantities.

Drying

The cost of drying the works would depend on time of year, type of job, geographical location etc. or alternatively a sum is provided for temporary operation of the Heating Installation. As this covers all labour, fuel and equipment, no further action is required by the estimator.

Cleaning

The extent of cleaning will depend upon the following:

(1) Alteration or new work—generally alteration works have to be more frequently cleaned due to tighter working space and avoidance of nuisance to occupiers.

(2) Type of finish, internal and external.

The main contractor should ensure that sub-contractors are required to deposit rubbish at a disposal point. He then allows for cleaning out own rubbish and for labour, lorry transport and tip charges in removal. Cleaning may be done either by working out labour costs or by obtaining quotations from commercial cleaning firms.
It is possible by experience and feedback to work out a rate per square metre of floor area for cleaning and to apply this in working out the sum to be allowed.

(6) Works by Nominated Sub-contractors

See Chapter 1.

(7) Goods and Materials from Nominated Suppliers

See Chapter 1.

(8) Contingencies

Most contracts for new work allow 3% for contingencies. An amount can be stated in the Bill of Quantities or a percentage added to the summary page or as shown in the first part of this chapter (*see* Dayworks) in an attempt to make this sum part of the competitive offer.

5 Excavations

HAND EXCAVATIONS

Schedule of Labour Constants for Hand Excavation

For normal soils the following items for excavating 1 m^3 of material may be taken as average.

	Hours
(1) Excavate and get out:	
(*a*) Surface excavation	1·75 to 2·00
(*b*) Vegetable soil	1·50
(*c*) Basements	2·00
(*d*) Foundation trenches	2·50
(*e*) Pits	3·00
(2) (*a*) Wheel n.e. 50 m, deposit, and return empty	0·75
(*b*) Every additional 50 m	0·75
(3) Spread and level	0·40
(4) Throw 1 stage of 1·50 m	1·00
(5) Re-excavate from spoil heaps	1·50
(6) Refill and ram	1·50
(7) Excavate through rock:	
(*a*) Basements, etc.	10·00
(*b*) Trenches	12·50

A point to remember about this trade is that excavated earth increases in bulk by 10 to 33⅓%; for excavating in sand the output should increase by about 33⅓%; and for excavating in heavy clay it should decrease by about the same amount.

Hand excavation today is normally only carried out on restricted sites, confined areas, etc., and the following items have been priced with such situations in mind.

Excavation is no longer measured in stages, or layers, of 1·50 m deep but in maximum (cumulative) depths of not exceeding 1·00 m, not exceeding 2·00 m and thereafter in cumulative stages of 4·00, 6·00 etc. m. This necessitates averaging the maximum depths where necessary in terms of 1·50 m stages as this is considered to be the maximum height for a throw where hand excavation is being carried out; and for each 1·50 m stage 1 hour per throw per m^3 is allowed. There are two methods of procedure. Method 1 is to ascertain the average amount of hours for the total depth. Method 2 is to price each 1·50 m stage and average the price.

Method 1

Assume a cube of ground 1 m long and 1 m wide to the depths shown.

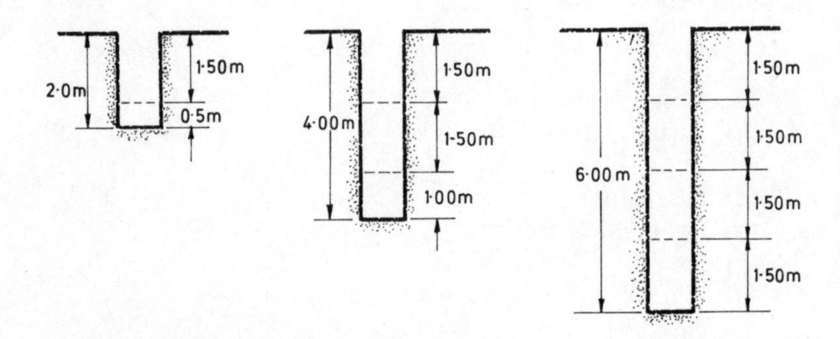

Trench Excavations

For a 2·00 m depth:

Excavate and get out; for a depth of 1·50 m is 1·50 m^3 @ 2·50 hours	3·75
Excavate and get out; from 1·50 to 2·00 m depth is 0·50 m^3 @ (2·50 + 1 hour for a throw) = 0·50 m^3 @ 3·50 hours	1·75
	5·50

∴ For 1·00 m^3 n.e. 2·00 m deep the average time is 5·50 ÷ 2 — 2·75 hours per m^3

For a 4·00 m depth:

Excavate and get out; for a depth of 1·50 m = 1·50 m^3 @ 2·50 hours	3·75
Excavate and get out; from 1·50 to 3·00 m depth is 1·50 m^3 @ (2·50 + 1 hour)	5·25
Excavate and get out; from 3·00 to 4·00 m depth is 1·00 m^3 @ (2·50 + 2 hours)	4·50
	13·50

∴ For 1·00 m^3 n.e. 4·00 m deep the average time is 13·50 ÷ 4 — 3·375 hours per m^3

For a 6·00 m depth:

Excavate and get out: for a depth of 1·50 m = 1·50 m^3 @ 2·50 hours	3·75
Excavate and get out; from 1·50 to 3·00 m depth is 1·50 m^3 @ (2·50 + 1 hour)	5·25
Excavate and get out; from 3·00 to 4·50 m depth is 1·50 m^3 @ (2·50 + 2 hours)	6·75
Excavate and get out; from 4·50 to 6·00 m depth is 1·50 m^3 @ (2·50 + 3 hours)	8·25
	24·00

∴ For 1·00 m^3 n.e. 6·00 m deep the average time is 24·00 ÷ 6 — 4·00 hours per m^3

By applying the same procedure to basements and pits whereby the labour constants are 2·00 hours and 3·00 hours per m^3 respectively, the average times are:

	Basements Hours per m^3	*Pits* Hours per m^3
For a 2·00 m depth	2·25	3·25
For a 4·00 m depth	2·875	3·875
For a 6·00 m depth	3·50	4·50

METHOD 2
(ILLUSTRATED BY EXAMPLES)

Example 1

Excavate foundation trenches, commencing below reduced level excavation and not exceeding 2·00 m deep. (This necessitates an additional throw from 1·50 to 2·00 m deep.)

Excavate and get out; (for 1st layer of 1·50 m) labourer 2·50 hours @ 190p	475p
Excavate and get out; (for 2nd layer of 1·50 m) labourer 2·50 + 1·00 (for one throw) = 3·50 hours @ 190p	665p
∴ 1·50 m^3 @ 475p	713p
and 0·50 m^3 @ 665p	333p
Cost of 2·00 m^3	1,046p
∴ Cost of 2·00 m^3 = 1,046 ÷ 2	523p
Profit and Oncost 20%	105p
Rate per m^3	£6·28

Example 2

Excavate foundation trenches commencing below reduced level excavation and not exceeding 4·00 m deep. (This necessitates two extra throws which includes for clearing back from side of trench.)

Excavate and get out; (for 1st layer of 1·50 m) labourer 2·50 hours @ 190p	475p
C/fd	475p

	B/fd
Excavate and get out; (for 2nd layer of 1·50 m) labourer 3·50 hours @ 190p	66[illegible]
Excavate and get out; (for 3rd layer of 1·50 m) labourer 2·50 + 2·00 (for two throws) = 4·50 hours @ 190p	855p
∴ 1·50 m³ @ 475p	713p
and 1·50 m³ @ 665p	998p
and 1·00 m³ @ 855p	855p
Cost of 4·00 m³	2,566p
∴ Cost of 1·00 m³ = 2,566 ÷ 4	642p
Profit and Oncost 20%	128p
Rate per m³	£7·70

Example 3

Excavate to reduce levels

Excavate and get out; labourer	1·75 hours	
Wheel n/e 50 m and deposit; labourer	0·75 hours	
	2·50 hours @ 190p	
		475p
Profit and Oncost 20%		95p
Rate per m³		£5·70

Note: Should the depth of excavation not exceed 225 mm deep a labour constant of 2·00 hours per m³ for excavation could be applied.

BY METHOD 1

Example 4

Excavate foundation trenches, commencing below reduced level excavation and not exceeding 1·00 m deep. (This includes throwing spoil to sides of trench.)

Excavate and get out; labourer 2·50 hours @ 190p	475p
Profit and Oncost 20%	95p
Rate per m^3	£5·70

Example 5

Excavate foundation trenches, commencing below reduced level excavation and not exceeding 2·00 m deep. (This necessitates an additional throw from a depth of 1·50 m to 2·00 m and includes for clearing back from edge of trench.)

Excavate and get out; for a depth of 1·50 m is 1·50 m^3 @ 2·50 hours	3·75
Excavate and get out; from 1·50 to 2·00 m depth is 0·50 m^3 @ 3·50 hours	1·75
	5·50

Average rate per m^3 = 5·50 ÷ 2 = 2·75 hours

∴ Labourer 2·75 hours @ 190p	523p
Profit and Oncost 20%	105p
Rate per m^3	£6·28

Example 6

Excavate foundation trenches commencing below reduced level excavation and not exceeding 4·00 m deep. (This necessitates two extra throws which include for clearing back from edge of trench.)

Excavate and get out; for a depth of 1·50 m is 1·50 m^3 @ 2·50 hours	3·75
Excavate and get out; from 1·50 to 3·00 m depth is 1·50 m^3 @ 3·50 hours	5·25
Excavate and get out; from 3·00 to 4·00 m depth as 1·00 m^3 @ 4·50 hours	4·50
	13·50

Average rate per m^3 = 13·50 ÷ 4 = 3·375 hours

∴ Labourer 3·375 hours @ 190p	641p
Profit and Oncost 20%	128p
Rate per m^3	£7·69

Example 7

Excavate foundation trenches commencing 3·00 m below ground level and not exceeding 1·00 m deep. It has been visualized that it would not be possible to make an earth ramp down to start of trench excavation level. (This necessitates one throw to the lower level, 3·00 m below ground level, which is included in the time allowed, and two additional throws from basement; i.e. one for each 1·50 m stage in depth, or part thereof.)

Excavate and get out;	2·50 hours
Two additional throws;	2·00 hours

Labourer 4·50 hours @ 190p	855p
Profit and Oncost 20%	171p
Rate per m^3	£10·26

Note: Using a ramp and wheeling would be about the same cost.

Example 8

Basement excavation commencing at ground level and not exceeding 2·00 m deep

Excavate and get out; for a depth of 1·50 m is 1·50 m^3 @ 2·00 hours	3·00
Excavate and get out; from 1·50 to 2·00 m depth is 0·50 m^3 @ 3·00 hours	1·50
	4·50

Average rate per m^3 = 4·50 ÷ 2 = 2·25 hours

∴ Labourer 2·25 hours @ 190p	428p
Profit and Oncost 20%	86p
Rate per m^3	£5·14

Example 9

Basement excavation commencing at ground level and not exceeding 4·00 m deep.

Excavate and get out; for a depth of 1·50 m is 1·50 m^3 @ 2·00 hours	3·00
Excavate and get out; from 1·50 to 3·00 m depth is 1·50 m^3 @ 3·00 hours	4·50
Excavate and get out; from 3·00 to 4·00 m depth is 1·00 m^3 @ 4·00 hours	4·00
	11·50

Average rate per m^3 = 11·50 ÷ 4 = 2·875 hours

∴ Labourer 2·875 hours @ 190p	546p
Profit and Oncost 20%	109p
Rate per m^3	£6·55

Example 10

Level and compact bottom of trench to receive concrete foundations.

Level and compact: Labourer 10 minutes

= $\frac{1}{6}$ hour @ 190p	32p
Profit and Oncost 20%	6p
Rate per m^2	38p

Example 11

Return, fill, and ram excavated materials in 300 mm layers to foundation trenches.

Re-excavating from spoil heaps beside trench, fill, and ram:

Labourer 1·50 hours @ 190p	285p
Profit and Oncost 20%	57p
Rate per m^3	£3·42

Example 12

Excavate vegetable soil average 500 mm deep, remove a distance not exceeding 50 lin.m, deposit, spread, and level.

Excavate and get out	1·50 hours
Wheel not exceeding 50 m	0·75 hour
Deposit, spread and level	0·40 hour
Total	2·65 hours

m^3 costs 2·65 x 190p = $503\frac{1}{2}$p

$\therefore$ Cost of 1 m^2 500 mm deep = $\frac{500}{1{,}000} \times 503\frac{1}{2}$p   252p

Profit and Oncost 20%	50p
Rate per m^2	3·02

Example 13

Spread and level excavations over area of site not exceeding 50 lin.m. from excavation.

There are two ways of pricing this item, depending on whether the estimator decides:

(*a*) That spreading and levelling can be carried out immediately, or

(*b*) that it will be necessary to have temporary spoil heaps.

Thus

(*a*) Wheel not exceeding

50 m	0·75 hours	
Spread and level	0·40 hours	
	1·15 hours @ 190p	
		219p
Profit and Oncost 20%		44p
Rate per m^3		£2·63

(*b*) Re-excavate from spoil heap at side of trench and fill barrows	1·50 hours	
Wheel not exceeding 50 m	0·75 hour	
Spread and level	0·40 hour	
	2·65 hours @ 190p	504p
Profit and Oncost 20%		101p
Rate per m³		£6·05

Example 14

Re-excavate from spoil heaps wheel materials not exceeding 100 lin.m. and deposit spread and level.

This item will only occur in a Bill of Quantities when the architect is not certain of his ground levels (e.g. in a housing scheme) and it is not at the estimator's discretion to indicate the treatment of excavated material.

Re-excavate from spoil heaps and fill into barrows	1·50 hours	
Wheel not exceeding 100 m	1·50 hours	
Deposit, spread and level	0·40 hour	
Labourer	3·40 hours @ 190p	646p
Profit and Oncost 20%		129p
Rate per m³		£7·75

Example 15

Remove excavations off site.

As explained in Example 13, there are two methods of dealing with removal of excavations:

(*a*) where the spoil produced by hand excavation is deposited in temporary spoil heaps adjacent to the excavations and re-excavated and loaded into lorry;

(*b*) where the spoil is loaded directly into the lorry as the excavations proceed, e.g. surface excavations.
(i) Assume tip to be 3 miles from site;
(ii) Tip dues to be £0·45 per load;
(iii) 5 tonne lorry which has a 3 m^3 capacity costs £4·00 per hour.

(*a*) Excavate from spoil heaps and load into lorry 3 m^3 @ 1·50 hours per m^3;

Labourer 4·50 hours @ 190p		855p
Lorry		
Standing time (6 men loading)		
$\frac{4\cdot50}{6}$	45 minutes	
Travelling to tip (12 miles per hour) $\frac{60}{12}$ x 3 miles	15 minutes	
Returning from tip	15 minutes	
Unloading at tip	10 minutes	
	85 minutes	
1 hour 25 minutes of 5 tonne lorry @ £4·00 per hour		567p
Tip dues 45p		45p
		1,467p
Allow 20% for increase in bulk of loose materials on 612p		122p
		1,589p
Profit and Oncost 20%		318p
Cost of disposing of 3 m^3 of loose spoil		1,907p

$\therefore$ Rate per m^3 = $\frac{1{,}907\text{p}}{3}$ = £6·36

(*b*) It is assumed that this case is surface excavation, excavated and loaded directly into lorry, and the rate for this part is included with surface excavations.
Surface excavation; labourer 2·00 hours per m^3,
$\therefore$ 3 m^3 requires 6 hours.

Lorry

Standing time (six men digging and loading)

$\frac{6 \cdot 00}{6} =$	60 minutes
Travelling to tip	15 minutes
Returning from tip	15 minutes
Unloading at tip	10 minutes
	100 minutes

1 hour 40 minutes of 5 tonne lorry @ £4·00 per hour	667p
Tip dues	45p
	712p
Allow 20% for increase in bulk of loose materials	142p
	854p
Profit and Oncost 20%	171p
Cost of disposal of 3 m³ of loose soil	1,025p
$\therefore$ Rate per m³ $= \frac{1{,}025p}{3} =$	£3·42p

Example 16

Extra over trench excavations for excavating in rock using wedges (no blasting etc. allowed).

The time required to excavate hard rock by hand is five times that for ordinary excavations. The time required for excavation in ordinary ground is deducted from the total time in order to derive the extra over rate.

Excavate in rock and get out:	
Labourer 12·50 hours @ 190p	2,375p
Deduct for do. in ordinary excavation:	
Labourer 2·50 hours @ 190p	475p
	1,900p
Profit and Oncost 20%	380p
Extra over rate per m³	£22·80p

Example 17

Extra over surface excavation for excavating through 150 mm thick bed of concrete. Assume hand labour.

A labourer will take about 1·50 hours per m² to break up and lay aside

Labourer 1·50 hours @ 190p	285p
Deduct cost of one m² of 150 mm thick reduced level digging	
Example 3 (1·75 hours @ 190p = 332½p)	
$332\frac{1}{2}\text{p} \times \frac{150}{1{,}000}$	50p
	235p
Profit and Oncost 20%	47p
Extra over rate per m²	£2·82

Note: The amount deducted depends upon which of the various descriptions embraces the extra over item (e.g. surface, basement, etc.).

Example 18

Earthwork support to sides of excavation not exceeding 2·00 m deep and not exceeding 2·00 m between opposing faces.

Consider one length of trench 50·00 m long, 0·75 m wide and 2·00 m deep; and assume firm ground having 200 x 38 mm poling boards at 2·00 m centres with two 100 x 100 mm struts between.

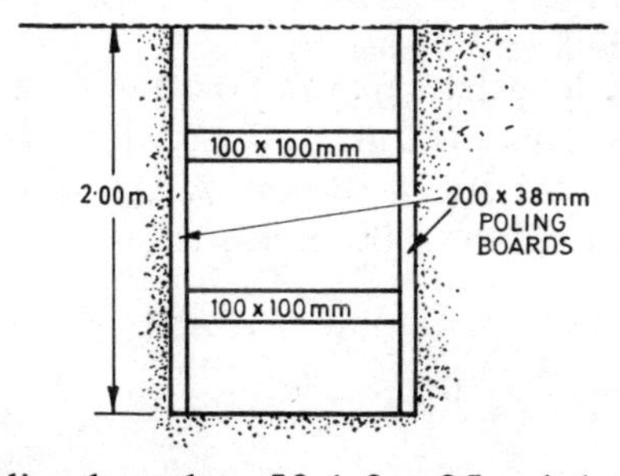

Number of poling boards = 50 ÷ 2 = 25 + 1 (at ends) = 26 x 2 (sides) = 52
Number of struts = 26 x 2 = 52

Quantity of timber required

Poling boards	52/0·200 x 0·038 x 2·00	= 0·79
Struts	52/0·67 x 0·100 x 0·100	= 0·35
		1·14 m^3

1·14 m^3 @ £108·00 per m^3 = £123·12

Assume 15 uses ∴ Cost per use = £123·12 ÷ 15 821p

(Timberman receives $1\frac{1}{2}$p per hour over labourer's rate-in-accordance with N.W.R.A.)

Timberman fixing and withdrawing after use:
0·14 m^3 per hour = (1·14 ÷ 0·14) x $191\frac{1}{2}$p 1,559p

2,380p

Area measured = 2/50·00 x 2·00 = 200 m^2

∴ Cost per m^2 = 2,380p ÷ 200	12p
Profit and Oncost 20%	$2\frac{1}{2}$p
Rate per m^2	$14\frac{1}{2}$p

Example 19

Earthwork support to sides of excavation of unstable ground not exceeding 2·00 m deep and not exceeding 2·00 m between opposing faces.

Consider one length of trench 50·00 m long, 0·75 m wide and 2·00 m deep; and assume 150 x 25 mm close sheeting is required having two 200 x 38 mm walings in the height and 100 x 100 mm struts at 2·00 m centres.

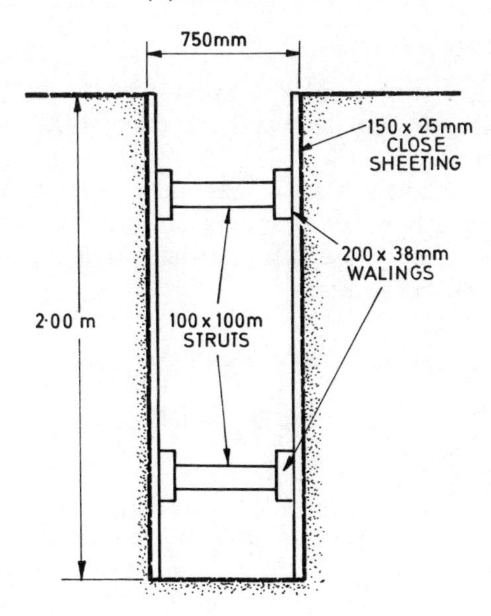

Number of struts = 50 ÷ 2 = 25 + 1 = 26 x 2 = 52

Quantity of timber required

Close sheeting	2/50·00 x 2·00 x 0·025 =	5·00
Walings	2/2/50·00 x 0·200 x 0·038 =	1·52
Struts	52/0·624 x 0·100 x 0·100 =	0·32
		6·84 m³

6·84 m³ @ £108·00 per m³ = £738·72

Assume 15 uses ∴ Cost per use = £738·72 ÷ 15	4,925p
Timberman fixing and withdrawing after use: 0·10 m³ per hour = (6·84 ÷ 0·10) x 191½p	13,099p
	18,024p

Area measured = 2/50·00 x 2·00 = 200 m²

∴ Cost per m² = 18,024p ÷ 200	90p
Profit and Oncost 20%	18p
Rate per m²	£1·08

Example 20

Earthwork support to sides of excavation of unstable ground not exceeding 4·00 m deep and exceeding 4·00 m between opposing faces.

Assume it is to the sides of basement construction and consider a length of 50·00 m long and average 3·00 m deep with 200 x 50 mm close sheeting and walings, 150 x 150 mm shores at 2·00 m centres.

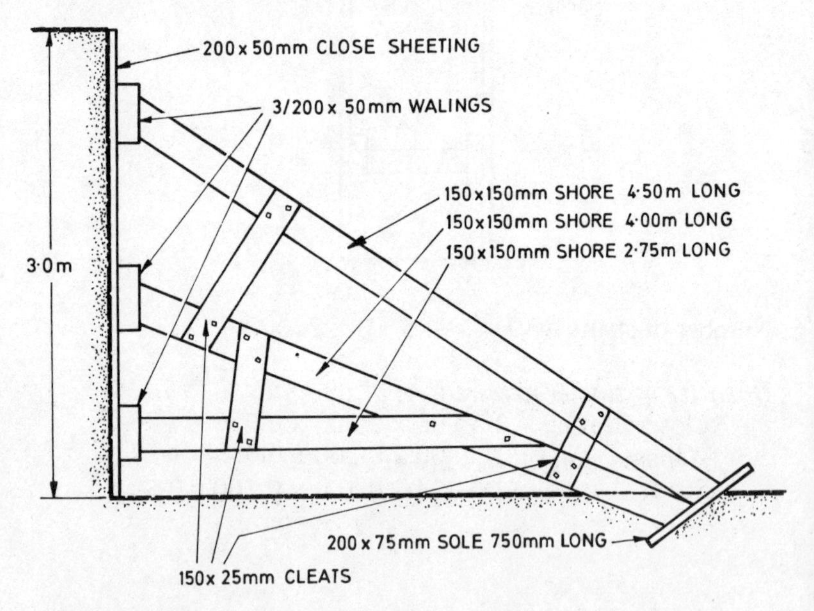

Number of shores required = 50 ÷ 2 = 25 + 1 = 26

Quantity of timber required

Close sheeting	50·00 x 3·00 x 0·050 = 7·50
Walings	3/50·00 x 0·200 x 0·050 = 1·50
Shores 4·50 + 4·00	
+ 2·75 = 11·25	∴ 26/11·25 x 0·150 x 0·150 = 6·58
Cleats allow	
2·00 m	∴ 26/2·00 x 0·150 x 0·025 = 1·17
Sole	26/0·750 x 0·200 x 0·075 = 0·29
	17·04 m^3

17·04 m^3 @ £108·00 per m^3 = £1,840·32

Assume 15 uses ∴ Cost per use = £1,840·32 ÷ 15	12,269p
Timberman fixing and withdrawing after use: 0·10 m^3 per hour = (17·04 ÷ 0·10) x 191½p	32,632p
	44,901p

Area measured = 50·00 x 3·00 = 150·00 m^2

∴ Cost per m^2 = 44,901p ÷ 150	299p
Profit and Oncost 20%	60p
Rate per m^2	£3·59

Example 21

Broken stone hardcore filling over area.

Cost of hardcore per tonne d/d site @ £2·20 1 m^2 hardcore weighs 1·60 tonnes @ £2·20	352p
Hardcore consolidates about 20% in compaction, therefore add 25%	88p
Labourer barrowing and filling 1·20 m^3 per hour (consolidated) @ 190p ∴ 190p ÷ 1·20	158p
Ramming hardcore in 300 mm layers with mechanical punner: 2 m^3 per hour costing 40p per hour for hire, ∴ 40p ÷ 2	20p
Labourer operating do. @ 190p per hour plus 1p in accordance with N.W.R.A., ∴ 191p ÷ 2	96p
	714p
Profit and Oncost 20%	143p
Rate per m^3	£8·57

Note: The cost of placing hardcore may vary considerably as the delivery lorry may conveniently tip the hardcore direct into the final position. However, on confined sites, where this is not possible the hardcore would have to be tipped and barrowed and filled by hand. The estimator must allow for whatever conditions prevail.

Example 22

Bed of hardcore 150 mm thick blinded on top with ashes.

Cost of 1 m^2 of hardcore including consolidation from previous analysis (352p + 88p)		
$440p \times \frac{150}{1,000}$		66p
1 m^3 of ash costs £2·00 d/d site =	200p	
Consolidation 20%, ∴ add 25% =	50p	
	250p	
Blinding layer will require 25 mm thickness		
$250p \times \frac{25}{1,000}$		7p
For hardcore in beds labour output is about 0·80 m^3 per hour		
∴ Cost of labourer per m^2 =		
$£1{\cdot}90 \times \frac{1{\cdot}00}{0{\cdot}80} \times \frac{150}{1,000}$		36p
Spreading ash 25 mm consolidated thickness: labourer 0·05 hours @ 190p		10p
Ramming hardcore with mechanical punner will ram about 8 m^2 per hour costing 40p per hour for hire ∴ 40p ÷ 8		5p
Labourer operating do. @ 191p per hour, ∴ 191p ÷ 8		24p
		148p
Profit and Oncost 20%		30p
Rate per m^2		£1·78p

Example 23

Bed of hardcore 100 mm thick blinded on top with ashes.

Cost of 1 m^2 of hardcore including consolidation from previous analysis $440p \times \frac{100}{1,000}$	44p
Blinding layer of ashes from previous analysis, $\therefore \frac{25}{1,000} \times 250p$	7p
Cost of labour laying hardcore = $190p \times \frac{1 \cdot 00}{0 \cdot 80} \times \frac{100}{1,000}$	24p
Spreading ash 25 mm consolidated thickness: labourer 0·05 hours @ 190p	10p
Ramming hardcore with mechanical punner from previous analysis	5p
Labour operating do. from previous analysis	24p
	114p
Profit and Oncost 20%	23p
	137p
Rate per m^2	£1·37

Pipe Trenches (Section W)

Unlike excavations under Section D pipe trenches are measured by the linear metre stating the average depth, therefore the cubic content must be calculated and related to excavating one m^3 plus the costs of grading the bottom of the trench, filling and ramming, spreading and levelling or removing residue, and planking and strutting to sides of trench.

Total depth of trench	*Suggested width of trench*	
	Up to 200 mm pipe	200 mm to 300 mm pipe
Not exceeding 1·00 m	550 mm	650 mm
Not exceeding 2·00 m	600 mm	750 mm
Not exceeding 4·00 m	750 mm	900 mm
Not exceeding 6·00 m	900 mm	1,000 mm

Although the S.M.M. states that depth-range shall be in increments of 2·00 m the following three examples show the cost of excavating trench, filling and ramming, and spreading and levelling surplus for three separate depth layers in terms of stages of throw of 1·50 m; i.e. not exceeding 1·50 m; 1·50–3·00 m; 3·00–4·50 m. The requisite volume of each depth layer is then added together as required to find the total cost of excavating and filling and ramming.

Example A

Depth layer up to 1·50 m

Excavate and get out	2·50 hours per m^3
Fill and ram, and deposit spread and level surplus	1·50 hours per m^3
	4·00 hours per m^3

∴ 4 hours of labourer @ 190p per hour

760p per m^3

Example B

Depth layer 1·50 to 3·00 m

Excavate and get out	2·50 hours per m^3
Extra throw from 3·00 to 1·50 m	1·00 hours per m^3
Fill and ram, and deposit spread and level surplus	1·50 hours per m^3
	5·00 hours per m^3

∴ 5 hours of labourer @ 190p per hour 950p per m^3

Example C

Depth layer 3·00 to 4·50

Excavate and get out, etc. all as 1·50 to 3·00 m layer	5·00 hours per m^3
Extra throw from 4·50 to 3·00 m	1·00 hour per m^3
	6·00 hours per m^3

∴ 6 hours of labourer @ 190p per hour   1,140p per m^3

Example 24

Excavate pipe trenches for pipes not exceeding 200 mm diameter, not exceeding 2·00 m and average 1·25 m deep, including grading bottom, earthwork support to sides, filling in, compacting, and disposing of surplus soil.

Volume of trench = 1·00 x 0·60 x 1·25 = 0·75 m^3 per m of trench

Cost of 0–1·50 layer = 760p per m^3	
Cost of trench per m = 760p x 0·75	570p
Grading bottom of trench 0·60 x 1·00 = 0·60 m^2 @ 10 minutes per m^2 @ 190p per hour	19p
Earthwork support to sides of trench 2/1·00 x 1·25 m = 2·50 m^2 @ 12p (*see* Example 18)	30p
	619p
Profit and Oncost 20%	124p
Rate per m	£7·43

Example 25

Excavate pipe trenches for pipes not exceeding 200 mm diameter, not exceeding 4·00 m and average 2·50 m deep including do.

Total volume of trench	
= 1·00 x 0·75 x 2·50	1·88 m^3
Volume of trench up to 1·50 mm deep	1·13 m^3
∴ Volume of trench exceeding 1·50 m but not exceeding 3·00 m deep	0·75 m^3
∴ Up to 1·50 m layer is 1·13 m^3 @ 760p per m^3	859p
and the 1·50–3·00 m layer is 0·75 m^3 950p per m^3	713p
Grading bottom of trench 0·75 x 1·00 m = 0·75 m^2 @ 10 mins per m^2 @ 190p per hour	24p
Earthwork support to sides of trench 2/1·00 x 2·50 m = 5·00 m^2 @ 12p (*see* Example 18)	60p
	1,656p
Profit and Oncost 20%	331p
Rate per m	£19·87

Example 26

Excavate pipe trenches for pipes not exceeding 200 mm diameter, not exceeding 4·00 m and averaging 3·50 m deep, including do.

Total volume of trench		
= 1·00 x 0·75 x 3·50		2·63 m^3
Volume of trench up to 1·50 m deep		
= 1·00 x 0·75 x 1·50	1·125 m^3	
Volume of trench exc. 1·50 m but n.e. 3·00 m deep		
= 1·00 x 0·75 x 1·50	1·125 m^3	
		2·25 m^3

Volume of trench exc. 3·00 m but n.e. 4·50 m deep	0·38 m³
∴ up to 1·50 m layer is 1·125 m³ @ 760p per m³	855p
1·50 to 3·00 layer is 1·125 m³ @ 950p per m³	1,069p
3·00 to 4·50 layer is 0·38 m³ @ 1,140p per m³	433p
Grading bottom of trench 0·75 x 1·00 m = 0·75 m² @ 10 minutes per m² @ 190p per hour, say	24p
Planking and strutting 2/1·00 x 3·50 m = 7·00 m² @ 54p say	378p
	2,759p
Profit and Oncost 20%	552p
	3,311p
Rate per m	£33·11

Note: Should a part of the excavations be required to be removed off site, say 10%, then in the above case the removal of 0·26 m³ should be added to the price thus:

Cost of excavating trench as above	2,759p
Remove excavations off site 0·26 m³ @ 854p (*see* Example 15(b))	222p
	2,981p
Profit and Oncost 20%	596p
	3,577p
Rate per m	£35·77

MACHINERY EXCAVATIONS

Example 27

Excavate over surface of site to reduce levels, wheel a distance not exceeding 100 m and deposit in permanent spoil heaps. 6,000 m^3

Note: (1) Average depth approximately 600 mm

(2) Conditions are suited to a tractor drawn scraper

(*a*) Hire rate per 40 hour week for Class 10 tractor £255·00

(*b*) Hire rate per 40 hour week for scraper of rated capacity 6 to 7·9 c. yd £85·00

£340·00

Cost of Machinery	£340·00	
Labour		
Driver 190p x 40 plus 180p	£77·80	
Banksman 190p x 40	£76·00	
Cost per Week	£493·80	
Cost per hour = £493·80 ÷ 37 = (*see* Chapter 3, Example 1(b))		1,335p
Cleaning and maintaining $\frac{1}{16}$ of $\frac{£77 \cdot 80}{40}$		12p
Fuel 18 litres @ 9p		162p
Insurance		3p
Cost of Machinery per hour		1,512p

Output

(4·5 m^3) tractor drawn scraper at 100 m per haul = 25 m^3 per hour	
∴ Cost per m^3 = 1,512p ÷ 25 =	61p
Cost of transport to and from site say £200·00	
∴ Rate per m^3 = £200·00 ÷ 6,000 =	3p
	64p
Profit and Oncost 20%	13p
	77p
Rate per m^3	77p

Example 28

(*a*) Basement excavations not exceeding 2·00 m deep
(*b*) Remove excavations off site. 600 m^3

It is decided to use a hired $\frac{3}{8}$ c. yd excavator, using dumpers to remove excavations directly to a tip 200 m from site. It is estimated that the excavator will take out approximately 13 m^3 per hour. There will be one man attending the excavator and one man at the tip due to special circumstances.

Preliminary Calculations
Output of excavator will be 1 m^3 every (60 ÷ 13) $4\frac{1}{2}$ minutes
Capacity of $2\frac{1}{2}$ c. yd dumper (1·91 m^3) is approximately 1·75 m^3 (allowing for bulking)
∴ time to load dumper = 1·75 x $4\frac{1}{2}$ = 8 minutes say
time to reach tip, dump and return = 8 minutes say

$$\text{Number of dumpers required} = \frac{\text{Time to load and haul}}{\text{Time to load}} = \frac{8+8}{8} = 2 \text{ dumpers}$$

Ideally the cost of the excavator should equal the cost of the removal equipment.

(*a*) Cost of excavator (from Chapter 3, Example 1(b))	876p
Output	
Excavator digs out 13 m^3 per hour, ∴ Cost per m^3 = 876p ÷ 13	67p
Transport	
To and from site, say £70·00 ÷ 600 m^3 =	12p
	79p
Profit and Oncost 20%	16p
	95p
Rate per m^3 is	95p

(*b*) Remove excavations off site.

Dumper

Hire of machines £92·00 plus 20% for road licence =	£110·40

Labour

Driver 190p x 40 plus 80p =	£ 76·80
Cost per week	£187·20

∴ Cost per hour = £187·20 ÷ 37 =	506p
Cleaning and maintenance $\frac{1}{16}$ of $\frac{£76{\cdot}80}{40}$	12p
Fuel 2·27 litres @ 15p per litre	34p
Insurance say	1p
Cost of two dumpers = 2 x 553p	1,106p
1 attendant at tip @ 190p per hour	190p
Cost per hour	1,296p

Output

The output is related to the excavator machine which has been taken as 13 m³ per hour, ∴ Cost per m³ = 1,296p ÷ 13 =	100p

Transport

As they are-road licensed; say driven under own power to and from owner's depot: 1 hour each way (553p – 12p maintenance) ∴ 2 x 541p = 1,082p ÷ 600 m³	2p
Local Dumping charge say 35p per load ∴ Cost per m³ = 35p ÷ 1·75 =	20p
	122p
Profit and Oncost 20%	24p
Rate per m³	£1·46

Example 29

Excavate pipe trenches for pipes not exceeding 200 mm diameter, not exceeding 4·00 m and average 2·50 m deep including grading bottom, earthwork support to sides, filling in and compacting, and disposing of surplus soil.

It is decided to use a $\frac{1}{4}$ c. yd (0·19 m^3) excavator with backacter.

Hire of machine for 40 hour week	£156·00
Labour	
Driver @ 190p x 40 plus £1·00	£ 77·00
Banksman 190p x 40	£ 76·00
	£309·00

Machine is used for 37 hours allowing 3 hours standing time per week	
∴ Cost per hour = £309·00 ÷ 37	835p
Cleaning and maintaining $\frac{1}{2}$ hour per 8 hour day = $\frac{1}{16}$ x 192$\frac{1}{2}$p	12p
Fuel 2·27 litres @ 9p per litre	20p
Insurance	1p
Oil, grease, etc. included	–
	868p

Output

$\frac{1}{4}$ c. yd backacter excavates 7·65 m^3 per hour using a backacter up to 2·00 m deep and 7·06 m^3 (7·65 m^3 less 7$\frac{1}{2}$%) from 2·00 to 4·00 m deep.

∴ Cost per m^3 up to 2·00 m deep = 868 ÷ 7·65 m^3	114p
Cost per m^3 from 2·00 m to 4·00 m deep = 868 ÷ 7·06 m^3	123p

Volume of trench	
1·00 x 0·75 x 2·50	1·88 m^3
Up to 2·00 deep	
1·00 x 0·75 x 2·00	1·50 m^3
	0·33 m^3

Up to 2·00 m deep layer is 1·50 m^3 @ 114p	171p
and 2·00 to 4·00 m deep layer is 0·33 m^3 @ 123p	41p
Grading bottom of trench 0·75 x 1·00 m = 0·75 m^2 @ 10 minutes per m^2 @ 190p per hour	24p
Earthwork support to sides of trench 2/1·00 x 2·50 = 5·00 m^2 @ 12p (*see* Example 18)	60p
Fill and ram and deposit spread and level surplus 1·50 hours per m^3 ∴ Cost per m = 1·88 x 1·50 x 190p	536p
	832p
Profit and Oncost 20%	166p
Rate per m	£9·98

As compared with £19·87 in hand excavations in Example 25.

Excavations in Rock and Concrete

This can be measured either full value or extra over the category of excavation in which it occurs.

Compressors

Table for breaking up concrete etc.

Thickness	*Compressor and 2 No. Drillers in m^2 per hour*
150 mm Concrete	4·00
225 mm Concrete	2·75
300 mm Concrete	2·00

150 mm Reinforced concrete	3·33
225 mm Reinforced concrete	2·00
300 mm reinforced concrete	1·25

Example 30

Break up reinforced concrete 225 mm thick

Hire rate for compressor for 40 hour week	£26·50
2 No. pneumatic breakers @ £7·00 per week	£14·00
	£40·50
Labour	
Drive 190p x 40 plus 60p =	£76·60
One driller* @ 190p plus 1½p x 40 =	£76·60
(Driver uses other drill)	
*No working rule allowance; but trade practice	£193·70
Sharpening tools: Allow	£3·75
Cost per week	£197·45
Cost per hour = £197·45 ÷ 37 =	534p
Allowance for replacing broken tools (hirer's responsibility)	3p
Fuel 4 litres @ 9p	36p
Insurance	2p
Cost per hour	575p

Note: Consumable stores such as drill rods, augers, bits included in hire.

Compressor with two drills breaks up 2 m² per hour, ∴ Cost per m² = 575p ÷ 2	288p
Transport costs (which depend upon quantity) say	3p
C/fd	291p

	B/fd	291p
Profit and Oncost 20%		58p
		349p
Rate per m²		£3·49

Example 31

Extra over basement excavations for breaking up rock.

Using compressor from Example 30, it will break up approximately 0·60 m³ per hour

∴ Cost per m³ = 575p ÷ 0·60	958p
Transport costs	3p
	961p
Deduct basement excavation price (Example 28(a))	79p
	882p
Profit and Oncost 20%	176p
	1,058p
Extra over rate per m³	£10·58

6 Concrete Work

The cost of concrete is assessed by volume and reduced to superficial and linear measurements where appropriate. The materials required in the preparation of concrete are Portland cement, sand, and aggregate and are sold by the tonne.

This necessitates the conversion of weight to volume of dry materials and is shown thus:

1·44 tonnes cement	= 1 m^3
1·52 tonnes sand	= 1 m^3
1·60 tonnes aggregate	= 1 m^3

It should be noted that the ratio of volume to weight may vary for sands and the different types of aggregates used in different parts of the country; therefore the estimator should check these for the areas in which his company operates. The conversion factors shown above are those which will be used for this section.

ALLOWANCES FOR SHRINKAGE IN CONCRETE

In the process of mixing concrete a reduction in bulk of the constituent materials takes place, due to the finer material filling the interstices of the coarse aggregate, the compaction and the addition of water. There are various methods of assessing allowances for shrinkage in concrete, three of which are illustrated.

(1) Is to assume an average shrinkage of 20% to allow for compaction of the coarse and fine aggregates; therefore 25% is added to the cost of the dry mix to compensate for this. The appropriate ratio of cement is added but is ignored for the purpose of volume calculations as it is assumed to be the adhesive for bonding the coarse and fine aggregates. The resultant total volume is therefore divided by the nominal volume of coarse and fine aggregates, thus:

Concrete 1:2:4

1 m^3 cement, i.e. 1·44 tonnes @ £24·00	3,456p
2 m^3 sand, i.e. 3·04 tonnes @ £2·10	638p
4 m^3 aggregate, i.e. 6·40 tonnes @ £2·50	1,600p
	5,694p
Shrinkage is 20%, ∴ add 25%	1,424p
Materials cost of 6 m^3	7,118p
∴ Materials cost of 1 m^3 = 7,118 ÷ 6	1,187p

Concrete 1:3:6

1 m^3 cement, i.e. 1·44 tonnes @ £24·00	3,456p
3 m^3 sand, i.e. 4·56 tonnes @ £2·10	958p
6 m^3 aggregate, i.e. 9·60 tonnes @ £2·50	2,400p
	6,814p
Shrinkage is 20%, ∴ add 25%	1,704p
Materials cost of 9 m^3	8,518p
∴ Materials cost of 1m^3 = 8,518p ÷ 9	947p

Concrete 1:4:8

1 m^3 cement, 1·44 tonnes @ £24·00	3,456p
4 m^3 sand, i.e. 6·08 tonnes @ £2·10	1,277p
8 m^3 aggregate, i.e. 12·80 tonnes @ £2·50	3,200p
	7,933p
Shrinkage is 20%, ∴ add 25%	1,983p
Materials cost of 12 m^3	9,916p
∴ Materials cost of 1 m^3 = 9,916p ÷ 12	826p

(2) Is to allow for a 40% increase over the total volume of cement, sand, and aggregate and divide by the total nominal volume of cement, sand, and aggregate. This assumes a shrinkage of about 28·6% *over all materials* and is shown thus:

Concrete 1:2:4

1 m^3 cement, i.e. 1·44 tonnes @ £24·00	3,456p
2 m^3 sand, i.e. 3·04 tonnes @ £2·10	638p
4 m^3 aggregate, i.e. 6·40 tonnes @ £2·50	1,600p
	5,694p
Allow for shrinkage, ∴ add 40%	2,278p
Materials cost of 7 m^3	7,972p
∴ Materials cost of 1 m^3 = 7,972p ÷ 7	1,139p

Concrete 1:3:6

1 m^3 cement, i.e. 1·44 tonnes @ £24·00	3,456p
3 m^3 sand, i.e. 4·56 tonnes @ £2·10	958p
6 m^3 aggregate, i.e. 9·60 tonnes @ £2·50	2,400p
	6,814p
Allow for shrinkage, ∴ add 40%	2,726p
Materials cost of 10 m^3	9,540p
∴ Materials cost of 1 m^3 = 9,540p ÷ 10	954p

Concrete 1:4:8

1 m^3 cement, i.e. 1·44 tonnes @ £24·00	3,456p
4 m^3 sand, i.e. 6·08 tonnes @ £2·10	1,277p
8 m^3 aggregate, i.e. 12·80 tonnes @ £2·50	3,200p
C/fd	7,933p

	B/fd	7,933p
		7,933p
Allow for shrinkage, ∴ add 40%		3,173p
Materials cost of 13 m^3		11,106p
∴ Materials cost of 1 m^3 = 4,808 ÷ 13		854p

(3) Is to allow for a 50% increase over the total volume otherwise as Method (2). This assumes a shrinkage of $33\frac{1}{3}\%$ *over all materials* and is shown thus:

Concrete 1:2:4

Materials as Method (2)	5,694p
Allow for shrinkage, ∴ add 50%	2,847p
Materials cost of 7 m^3	8,541p
∴ Materials cost of 1 m^3 = 8,541p ÷ 7	1,220p

Concrete 1:3:6

Materials as Method (2)	6,814p
Allow for shrinkage, ∴ add 50%	3,407p
Materials cost of 10 m^3	10,221p
∴ Materials cost of 1 m^3 = 10,221 ÷ 10	1,022p

Concrete 1:4:8

Materials as Method (2)	7,933p
Allow for shrinkage, ∴ add 50%	3,966p
Materials cost of 13 m^3	11,899p
∴ Materials cost of 1 m^3 = 11,899p ÷ 13	915p

(4) For 'all-in' aggregate, i.e. aggregate containing sand which is used for low-grade concrete work, an addition of 15% should be made to the volume to allow for shrinking and compaction and the total quantity of 'all-in' aggregate and cement should then be divided by the nominal quantity of aggregate, e.g. for 1:12 divide by 12.

AMOUNT OF MATERIALS

The differing costs of materials in the three methods illustrated are accounted for as shown in the following Table.

Amount of Materials Required According to Mix

	Concrete mix	Total of materials plus shrinkage allowance	Materials reduced	Amount of materials required for 1 m³
Method (1)	1:2:4	7 + 25% = 8·75	8·75 ÷ 6	1·46
	1:3:6	10 + 25% = 12·50	12·50 ÷ 9	1·39
	1:4:8	13 + 25% = 16·25	16·25 ÷ 12	1·35
Method (2)	1:2:4	7 + 40% = 9·8	9·8 ÷ 7	1·40
	1:3:6	10 + 40% = 14·00	14·00 ÷ 10	1·40
	1:4:8	13 + 40% = 18·2	18·2 ÷ 13	1·40
Method (3)	1:2:4	7 + 50% = 10·5	10·5 ÷ 7	1·50
	1:3:6	10 + 50% = 15	15·00 ÷ 10	1·50
	1:4:8	13 + 50% = 19·5	19·5 ÷ 13	1·50

C.P. 110: Part 1: 1972, Table 50, gives recommendations for Standard Mixes which are determined by weight. These weights would be determined by using a weigh batcher and reconciling the quantities to the required volume.

Note on the Methods of Specifying Concrete: BS 5328: 1976

1. This standard gives methods of specification for concrete as a site construction material either as ready-mixed or mixed on site.
   It takes into account the client, the contractor and the producer and is thus suitable for purchase and supply.
   Twenty-four relevant British Standards are referred to in it mainly covering the constituent materials plus BS 1881, Methods of Testing Concrete.

2. *Types of Mix*

   2.1 Prescribed mixes – for which the purchaser specifies the proportions of the constituents and is responsible for ensuring that the proportions prescribed will produce a concrete with the performance he requires. Strength testing will not be used to judge compliance. (These mixes cover concreting needs in all but the most advanced engineering structures.)

   2.2 *Designed mixes* – for which the purchaser is responsible for specifying the required performance and the producer is responsible for selecting the mix proportions to produce the required performance. Strength testing will form an essential part of the judgement of compliance.
   (A designed mix to produce the qualities needed will usually be more economical but it places the onus on the producer of providing concrete of a given strength measured at a given age.)

N.B. The examples deal with prescribed mixes only. The required performance or proportions of the constituents are normally defined in the Preamble to the Bill of Quantities but as this book does not cover Preambles the required proportions of the constituents are defined in the examples given.

Method (1) is considered to be more accurate for estimating purposes and will be used in subsequent examples.

In the pricing of concrete there are three main points to be considered: cost of (1) materials; (2) mixing; (3) transporting, placing in position, and tamping, etc.

Concrete is normally mixed by machine and hand-mixing would only be used where the quantity of concrete required is minimal.

For machine-mixing the cost of a concrete mixer must be known. This cost may be based on the weekly rate of hire of the machine, but as most builders own their own concrete mixers the cost per hour is normally assessed on the average life of the machine.

There are various ways of assessing this cost; the most straightforward method is the Straight Line method and is shown thus:

(*x*) Cost of a 10/7 drum mixer when new — £2,200
less scrap value, say — £ 200

£2,000

Assume a 5-year life ∴ Annual cost =
2,000 ÷ 5 = — £ 400

∴ The cost per hour based on a usage of
1,500 hours per annum = 400 ÷ 1,500 — 27p

(*y*) Cost of a 7/5 drum mixer when new — £725
less scrap value, say — £ 75

£650

Assume a 5-year life ∴ Annual cost
= 650 ÷ 5 — £130

∴ The cost per hour based on a usage of
1,500 hours per annum = £130 ÷ 1,500 — 9p

AVERAGE OUTPUT OF CONCRETE MIXERS

The number of men required to operate a mixer varies according to the type of mixer and the size of the job. The aim is to achieve continuous production geared to the output of the placing squad.

Size		Output per hour m^3	Number of men attending
Imperial (c. ft)	Metric equivalent (m^3)		
5/3½	0·140/0·100	1·00	2
7/5	0·200/0·140	1·40	2
10/7	0·280/0·200	2·00	3
14/10	0·400/0·280	2·80	4

Note: Output of 5/3½ in Brickwork section taken as 1·15 m^3 per hour. This is because there are only two ingredients in mortar as against three in concrete.

0·200/0·140 means 0·200 m^3 input of dry materials for an output of 0·140 m^3 of mixed concrete, and applies to the other ratios shown. The 5/3$\frac{1}{2}$ (0·140/0·100) and 7/5 (0·200/0·140) are normally used for mortar mixing. The 7/5 (0·200/0·140), 10/7 (0·280/0·200), and the 14/10 (0·400/0·280) are normally used for concrete mixing.

TRANSPORTING

Barrows

Concrete transported by ordinary barrows takes $\frac{3}{4}$ to 1 hour per man per m^3 for a distance not exceeding 25 m.

Dumpers

Travelling time per load from mixing point to off-loading point, dumping and returning for next load:

Linear metres	Travelling time
100	0·11 hours
200	0·13 hours
300	0·16 hours
400	0·19 hours
500	0·22 hours

Dumpers have a capacity of 20% below their normal capacity for wet materials such as concrete.

The following table shows average times for transporting mixed concrete from the mixing point and placing in position. The figures quoted include for transporting a distance of not exceeding 25 m and also uplifting (by crane or otherwise) to various floor levels in the building. It should be noted that the cost of the crane is not included in the rates for concrete but would be included in the Preliminaries.

Transporting Times for Mixed Concrete

Material	Position	Transporting and Placing Labourer's Time in hours per m^3
Plain concrete	Foundations n.e. 100 mm thick	2·75
	Foundations 100 to 150 mm thick	2·5
	Foundations 150 to 300 mm thick	2
	Foundations over 300 mm thick	1·5
	Mass concrete	1·5
	Beds n.e. 100 mm thick	5·5
	Beds 100 to 150 mm thick	5
	Beds 150 to 300 mm thick	4
	Beds over 300 mm thick	3·25
Reinforced concrete	Suspended floor or roof slabs n.e. 100 mm thick	7·75
	Suspended floor or roof slabs 100 to 150 mm thick	7·5
	Suspended floor or roof slabs 150 to 300 mm thick	7
	Suspended floor or roof slabs over 300 mm thick	7
	Walls n.e. 100 mm thick	9
	Walls 100 to 150 mm thick	8
	Walls 150 to 300 mm thick	7
	Walls over 300 mm thick	7
	Deep beams n.e. 0·03 m^2 sectional area	9
	Deep beams 0·03 to 0·10 m^2 sectional area	8
	Deep beams 0·10 to 0·25 m^2 sectional area	7
	Deep beams over 0·25 m^2 sectional area	7
	Columns n.e. 0·03 m^2 sectional area	10
	Columns 0·03 to 0·10 m^2 sectional area	9
	Columns 0·10 to 0·25 m^2 sectional area	8·5
	Columns over 0·25 m^2 sectional area	8·5
	Isolated beam casings, lintels, etc.	10

HAND-MIXED CONCRETE

Example 1

Concrete (1:3:6) in foundations exceeding 100 mm but not exceeding 150 mm thick (20 mm aggregate).

Cement d/d site per tonne		2,400p
Unload $1\frac{1}{4}$ tonnes per hour of labourer @ 190p		152p
		2,552p
Materials per m^3		
Cement 1·44 tonnes @ 2,552p per tonne		3,675p
Concrete sand 1·52 tonnes @ 210p per tonne		319p
20 mm whin aggregate 1·60 tonnes @ 250p per tonne		400p
Mix by volume		
1 m^3 cement @ 3,675p per m^3	3,675p	
3 m^3 sand @ 319p per m^3	957p	
6 m^3 aggregate @ 400p per m^3	2,400p	
	7,032p	
Shrinkage is 20% ∴ add 25%	1,758p	
Materials cost for 9 m^3	8,790p	
∴ Materials cost for 1 m^3 =		977p
Labour hand mixing	5·0	
Labour transporting and placing	2·5	
	7·5 hours @ 190p	1,425p
		2,402p
Profit and Oncost 20%		480p
		2,882p
Rate per m^3		£28·82

Example 2

Concrete (1:3:6) in foundations over 300 mm thick (20 mm aggregate)

Materials cost for 1 m³ (as Example 1)		977p
Labour hand-mixing	5·0 hours	
Labour placing	1·5 hours	
	6·5 hours	
	@ 190p	1,235p
		2,212p
Profit and Oncost 20%		442p
		2,654p
Rate per m³		£26·54

MACHINE-MIXED CONCRETE

The working cost of the mixer may be calculated prior to the building up of a rate, or it may be calculated with the item.

(*a*) *Working cost per hour of* 7/5 (0·200/0·140) *mixer*

Cost per hour (from Straight Line method (*y*))	9p
General maintenance, repairs, etc. 10%	1p
Diesel fuel 1·50 litres of gas oil @ 9p	14p
Oil and grease	3p
Insurance	2p
	29p

Labour

1 labourer operating and filling 190p plus 1p =	191p	
1 labourer filling 190p	190p	381p
	C/fd	410p

	B/fd 410p	
Cleaning and maintaining machine ½ hour per 8-hour day; ∴ $\frac{1}{16}$ of 191p	12p	
Cost of mixer and labour per hour	422p	
Output per hour = 1·40 m³ ∴ Cost of mixer and labour per m³ = 422p ÷ 1·40		302p

(*b*) *Working cost per hour of* 10/7 (0·280/0·200) *mixer*

Cost per hour (from Straight Line method (*x*))		27p
General maintenance, repairs etc. 10%		3p
Diesel fuel 1·80 litres @ 9p		16p
Oil and grease		4p
Insurance		3p

Labour

1 Labourer operating and filling 190p plus 1p =	191p	
2 labourers filling @ 190p	380p	571p
Cleaning and maintaining machine ½ hour per 8-hour day; ∴ $\frac{1}{16}$ of 191p		12p
Cost of mixer and labourer per hour		636p
Output per hour = 2·00 m³ ∴ Cost of mixer and labour per m³ = 636p ÷ 2		318p

(*c*) *Working cost per hour of* 14/10 (0·400/0·280) *mixer*

Cost per hour say	53p
General maintenance, repairs, etc. 10%	5p
C/fd	58p

B/fd		58p
Diesel fuel 2·27 litres @ 9p		21p
Oil and grease		5p
Insurance		4p
		88p

Labour

One labourer operating and filling 190p plus 1p =	191p	
Three labourers filling @ 190p	570p	761p
		849p
Cleaning and maintaining machine $\frac{1}{2}$ hour per 8-hour day $\therefore \frac{1}{16}$ of 191p		12p
Cost of mixer and labour per hour		861p

Output per hour = 2·80 m^3
∴ Cost of mixer and labourer per m^3 =
861p ÷ 2·80 — 308p

The following example is set out in detail for the benefit of students taking examinations. Subsequent examples will not repeat detail.

Example 3

Concrete (1:3:6) in foundations exceeding 100 mm but not exceeding 150 mm thick (20 mm aggregate)

Cement d/d site per tonne	2,400p
Unload $1\frac{1}{4}$ tonnes per hour of labourer @ 190p	152p
	2,552p

Materials per m^3

Cement 1·44 tonnes @ 2,552p per tonne	3,675p
Concrete sand 1·52 tonnes @ 210p per tonne	319p

20 mm whin aggregate 1·60 tonnes @ 250p per tonne			400p

Mix by Volume

1 m³ cement @ 3,675p	3,675p
3 m³ sand @ 319p per m³	957p
6 m³ aggregate @ 400p per m³	2,400p
	7,032p
Shrinkage is 20%, ∴ add 25%	1,758p
Materials cost for 9 m³	8,790p
∴ Materials cost per 1 m³ =	977p

Mixing

Using 10/7 (0·280/0·200) mixer	
Cost per hour is	27p
General maintenance, repairs, etc. 10%	3p
Diesel fuel 1·80 litres @ 9p	16p
Oil and grease	4p
Insurance	3p
	53p

Labour

1 labourer operating and filling 190p plus 1p =	191p	
2 labourers filling @ 190p	380p	571p
Cleaning and maintaining ½ hour per 8-hour day:		624p
∴ $\frac{1}{16}$ of 191p		12p
Cost of mixer and machine per hour		636p

C/fd	977p

B/fd	977p
Output per hour = 2·00 m^3	
∴ Cost of mixer and labour per m^3 = 636p ÷ 2	318p
Labour transporting and placing 2·5 hours @ 191p	478p
	1,773p
Profit and Oncost 20%	355p
	2,128p
Rate per m^3	£21·28

Example 4

Concrete (1:2:4) beds exceeding 100 mm but not exceeding 150 mm thick

Mix by volume	
1 m^3 cement @ 3,675p per m^3	3,675p
2 m^3 sand @ 319p per m^3	638p
4 m^3 aggregate @ 400p per m^3	1,600p
	5,913p
Shrinkage is 20%, ∴ add 25%	1,478p
Materials cost of 6 m^3	7,391p
∴ Materials cost for 1 m^3	1,232p
Mixing, using 7/5 mixer	302p
Transporting and placing 5 hours @ 190p	950p
	2,484p
Profit and Oncost 20%	497p
Rate per m^3	2,981p
∴ Rate per m^3	£29·81

See Example 20 at end of chapter for a similar example of prescribed mix by weight in accordance with C.P. 110: Part 1: 1972, Table 50 and B.S. 5328: 1976.

Example 5

Concrete (1:2:4) in horizontal suspended floors exceeding 150 mm but not exceeding 150 mm thick

Materials	1,232p
Mixing, using 10/7 mixer	318p
Transporting and placing 7 hours @ 190p	1,330p
	2,880p
Profit and Oncost 20%	576p
	3,456p
Rate per m^3	£34·56

Example 6

Concrete (1:2:4) in deep beams over 0·10 m^2 but not exceeding 0·25 m^2 in cross-sectional area

Materials	1,232p
Mixing, using 10/7 mixer	318p
Transporting and placing 7 hours @ 190p	1,330p
	2,880p
Profit and Oncost 20%	576p
	3,456p
Rate per m^3	£34·56

Example 7

Concrete (1:2:4) in columns over 0·10 m^2 but not exceeding 0·25 m^2 in cross-sectional area

Materials	1,232p
Mixing, using 10/7 mixer	318p
C/fd	1,550p

B/fd	1,550p
Transporting and placing 8·5 hours @ 190p	1,615p
	3,165p
Profit and Oncost 20%	633p
	3,798p
Rate per m^3	£37·98

Example 8

Hollow block suspended horizontal floor total thickness 150 mm with 300 x 300 x 100 mm hollow clay blocks at 375 mm centres and concrete filling to 75 mm ribs and 50 mm topping including 300 x 75 x 22 mm keyed slip tiles.

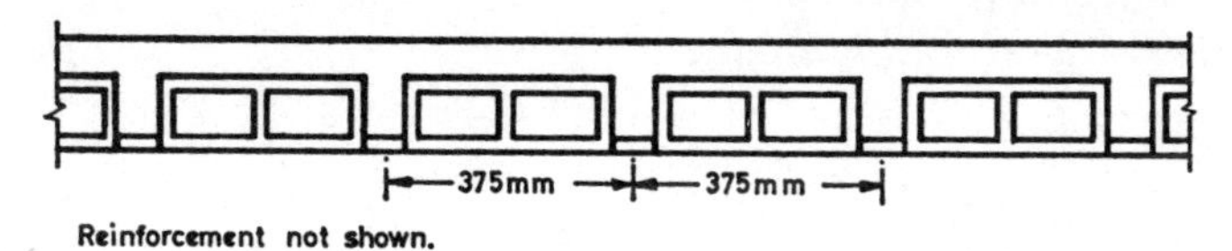

Reinforcement not shown.

Effective area of one block 375 mm x 300 mm

$$\therefore \text{ of blocks per m}^2 \ \frac{1{,}000 \text{ mm} \times 1{,}000 \text{ mm}}{375 \text{ mm} \times 300 \text{ mm}} = 8{\cdot}88$$

say 9

∴ 9 hollow blocks @ $11\frac{1}{2}$p	$103\frac{1}{2}$p	
and 9 slip tiles @ $10\frac{1}{4}$p	$92\frac{1}{4}$p	
Unloading 4 minutes labourer @ 190p (based on unloading average 300 per hour of the above)	$12\frac{3}{4}$p	
	$208\frac{1}{2}$p	
Allow 5% waste	$10\frac{1}{2}$p	219p
	C/fd	219p

B/fd 219p

Concrete to floor 1 m^2 (1:2:4)
50 mm thick = 0·050 m^3
Concrete to rib
3/1·00 x 0·08 x 0·08 = 0·019 m^3

0·069 m^3

(*see* Example 5) say 0·07 m^3 @ 1,550p — 109p

Labour
Hoisting and placing blocks and slip tiles in position
Labourer 1 hour @ 190p — 190p
Hoisting and placing concrete
to floor 0·05 m^3 @ 7·75 hours
per m^3 = 0·39
Hoisting and placing concrete
to rib 0·02 m^3 @ 9 hours per m^3
= 0·18

0·57 hours @
190p — 108p

626p
Profit and Oncost 20% — 125p

751p

Rate per m^2 — £7·51

STEEL REINFORCEMENT

Generally:

(1) Allow $2\frac{1}{2}$% waste on the cost of rods.
(2) Allow $2\frac{1}{2}$% for rolling margin on rods.
(3) Qualified bar benders and steel fixers receive tradesman's rate in accordance with the N.W.R.A.
(4) Allow an average of 15% for laps and 5% for waste on fabric reinforcement.

Rod Reinforcement

Labour Constants

Location of Work	Steel fixer: unloading, cutting, hooking, bending, assembling and placing Hours per 100 kg of mild steel rods			
	6 to 8 mm	10 to 12 mm	16 to 20 mm	25 mm
Foundations and beds	7	6	5	4
Floors and roofs	12	10	8	6
Walls	14	12	10	8
Beams, columns, staircases braces, cantilevers, etc.	16	14	12	10

Note: Where rods are supplied already cut, bent and labelled, the above figures can be reduced by $33\frac{1}{3}\%$.

Fabric Reinforcement

Labour Constants

Location of Work	Labourer unloading, cutting, laying, and tying with wire	
	Hours	Unit
Beds, landings, floors, roofs	$\frac{1}{4}$	m^2
Foundations	$\frac{1}{6}$	m^2
Walls	$\frac{1}{3}$	m^2
Staircases	$\frac{1}{2}$	m^2
Raking Cutting	$\frac{1}{4}$	m
Circular Cutting	$\frac{3}{8}$	m

Note: Fabric reinforcement is normally fixed by labourers.

Example 9

25 mm diameter mild steel rods in beams and columns, including all bends, hooks, tying wire, distance blocks, and spacers

100 kg of 25 mm rods @ £144·00 per tonne	1,440p
Add waste $2\frac{1}{2}\%$	36p
Add rolling margin $2\frac{1}{2}\%$	36p
C/fd	1,512p

	B/fd	1,512p
Spacers say		35p
Tying wire 0·66 kg @ 40p per kg		27p
Unloading, cutting, bending, and fixing		
10 hours steelfixer @ 220p		2,200p
		3,774p
Profit and Oncost 20%		755p
Rate per 100 kg		4,529p
∴ Rate per tonne		£452·90

Example 10

12 mm diameter mild steel rods in do. including do.

100 kg of 12 mm rods @ £165·00 per tonne	1,650p
Add waste $2\frac{1}{2}$%	41p
Add rolling margin $2\frac{1}{2}$%	41p
Spacers say	35p
Tying wire 1 kg @ 40p per kg	40p
Unloading, cutting, bending	
14 hours steelfixer @ 220p	3,080p
	4,887p
Profit and Oncost 20%	977p
Rate per 100 kg	5,864p
∴ Rate per tonne	£586·40

Example 11

6 mm diameter mild steel rods in walls including do.

100 kg of 6 mm rods @ £171·00 per tonne		1,710p
Add waste $2\frac{1}{2}$%		43p
Add rolling margin $2\frac{1}{2}$%		43p
Spacers say		48p
Tying wire 2 kg @ 40p per kg		80p
	C/fd	1,924p

	B/fd	1,924p
Unloading, cutting, bending 14 hours steelfixer @ 220p		3,080p
		5,004p
Profit and Oncost 20%		1,001p
Rate per 100 kg		6,005p
∴ Rate per tonne		£600·50

Example 12

C.283 long mesh fabric laid in floors, weighing 2·61 kg/m^2

Cost of 1 m^2 of long mesh fabric	59p
Allowance for laps 15%	9p
Allowance for waste 5%	3p
Tying wire and spacers (cost negligible)	3p
Labourer laying 4 m^2 per hour @ 190p	48p
	122p
Profit and Oncost 20%	24p
	146p
Rate per m^2	£1·46

FORMWORK

The timber used for formwork at the present time is mainly heavy plywood, 19 mm and 25 mm thick dependent on span for floor soffits, sides of columns, and sides and soffits of beams, the supports and struts being timber. Acrow props are also used in conjunction with timber as supports.

The amount of strutting timbers will vary according to the height of floors and beam soffits above floor level. The S.M.M allows for this by grouping soffit heights over 3·50 m high into further stages of 1·50 m; e.g. over 3·50 m and not exceeding 5·00 m high and so on.

Timber for formwork can be used about 3 or 4 times for columns and beams and about 5 times for floor soffits before becoming unusable.

Carpenters are entitled to an allowance of 1p per hour for re-use of old formwork and this normally includes preparing the new formwork.

Example 13

Formwork to horizontal soffit of concrete slabs.

Note: In accordance with the item the supports do not exceed 3·50 m high.

1 m^2 19 mm plywood @ £4·00 per m^2	400p
Timber in supports requires about 0·05 m^3 per m^2 of soffit	
∴ 0·05 m^3 softwood @ £108·00 per m^3	540p
Cost of timber	940p
Based on 5 uses, the cost per use is 940p ÷ 5	188p
Waste on timber per use, say 10%	19p
Allow for mould oil, nails, and bolts	24p
Labour making, fixing, and stripping timber	
Carpenter 1·5 hours @ 221p	332p
	563p
Profit and Oncost 20%	113p
	676p
Rate per m^2	£6·76

Alternative to Example 13

Using Acrow props or similar

1 m^2 19 mm plywood @ £4·00 per m^2	400p
Timber in supports requires 0·03 m^2 per m^2 of soffit	
∴ 0·03 m^2 softwood @ £108·00 per m^3	324p
Cost of timber	724p
Based on 5 uses, the cost per use is 724p ÷ 5	145p
Waste on timber per use, say 10%	15p
C/fd	160p

B/fd	160p
Acrow props hired @ 16p per week	
8 Acrow props are required per 13·50 m^2 of soffit	
∴ 4 weeks cost 8 x 16p x 4 = 512p	
∴ Cost per m^2 = $\frac{512}{13{\cdot}5}$	38p
Allow for mould oil, nails and bolts	24p
Labour making fixing and stripping	
Carpenter 1·5 hours @ 221p	332p
	554p
Profit and Oncost 20%	111p
	665p
Rate per m^2	£6·65

Example 14

Formwork to sides and soffits of horizontal beams

1 m^2 19 mm plywood @ £4·00 per m^2	400p
Timber in supports requires 0·06 m^3 per m^2 of beam	
∴ 0·06 m^3 softwood @ £108·00 per m^3	648p
Cost of timber	1,048p
Based on 3 uses, the cost per use is 1,048p ÷ 3	349p
Waste on timber per use, say 10%	35p
Allow for mould oil, nails, and bolts	24p
Labour making, fixing and stripping timber	
Carpenter 2·10 hours @ 221p	464p
	872p
Profit and Oncost 20%	174p
	1,046p
Rate per m^2	£10·46

From the rate per m² a rate per m, stating the girth, may be derived; e.g. 800 mm girth = $\frac{800}{1,000} \times 1{,}046\text{p} = £8{\cdot}37$

Example 15

Formwork to sides of columns

1 m² 19 mm plywood @ £4·00 per m²	400p
Timber in supports requires 0·03 m³ per m² of columns	
∴ 0·03 m³ softwood @ £1·08 per m³	324p
Cost of timber	724p
Based on 3 uses, the cost per use is 724p ÷ 3	241p
Waste on timber per use, say 10%	24p
Allow for mould oil, nails, and bolts	24p
Labour making, fixing, and stripping timber	
Carpenter 2·10 hours @ 221p	464p
	753p
Profit and Oncost 20%	151p
	904p
Rate per m²	£9·04

From the rate per m² a rate per m stating the girth, may be derived; e.g. 900 mm girth = $\frac{900}{1,000} \times 904\text{p} = £8{\cdot}14$

PRECAST CONCRETE

Example 16

100 x 250 mm rectangular lintel, left rough, reinforced with two 12 mm diameter mild steel rods.

Note: Concrete is specified to be (1:2:4) and the precast is being prepared on site by contractor.

A 100 x 250 mm lintel 2 m long requires 0·05 m³ concrete. Therefore 20 of such lintels require 1 m³ concrete. Therefore consider 20 lintels.

1 m³ concrete costs (*see* Example 5)		1,550p
9 hours labourer placing (as for beams) @ 190p		1,710p
a 2 m long lintel requires rods 2·30 m long;		
i.e. 4·60 m of rod for 20 lintels = 92 m of rod		
@ 0·888 kg per m = 81·70 kg		
∴ 81·70 kg @ 49p (*see* Example 10)		4,003p
		7,263p
A 2 m long lintel will require 1·27 m² of boarding including ends.		
∴ 1·27 m² 19 mm plywood @ £4·00	508p	
Mould oil, screws, and bolts	24p	
Mould clamps, say	8p	
Joiner making box, 1 hour @ 220p	220p	
Cost of one box	760p	
Allowing for 20 uses per box, the cost per use = $\frac{760}{20}$p = 38p		
∴ for 20 uses x 20 lintels = 38p x 20		760p
∴ Cost per m³ (or cost of casting 20 (2 m long lintels)		8,023p

The overall cost per m³ having been found, the cost of any rough rectangular or shaped lintel may be calculated on the basis of a ratio to 1 m² of sectional area 1 m long.

i.e. Cost of 100 x 250 mm lintel 1 m long		
$= 8{,}023\text{p} \times \frac{100 \times 250}{1{,}000 \times 1{,}000} =$		201p
Allow 2½% Waste		5p
The cost of stacking or unloading lintel:		
2 labourers 5 minutes each = 10 minutes		
@ 190p	32p	
Cost of taking lintel from stack and hoisting onto scaffold:		
C/fd	32p	206p

B/fd	32p	206p
2 labourers $7\frac{1}{2}$ minutes each = 15 minutes @ 190p	48p	
Cost of bricklayer and labourer lifting lintel from scaffold and set and level in position: 1 bricklayer and 1 labourer 10 minutes each = 10 minutes @ 410p	68p	148p
		354p
Profit and Oncost 20%		71p
		425p
Rate per m		£4·25

Example 17

100 mm x 150 mm rectangular lintel, left rough for plaster, reinforced with one 12 mm diameter mild steel rod.

(Assume the lintels are purchased from a precast-concrete manufacturer.)

Cost of 100 x 150 mm lintel per m £1·48	148p
Allow for Waste $2\frac{1}{2}$%	4p
From the previous example a comparative rate for unloading, hoisting on to scaffold, and setting in position may be deduced; i.e. total cost for setting 100 x 250 mm lintel = 148p	
∴ Cost for setting a 100 x 150 mm lintel = $\frac{100 \times 150}{100 \times 250} \times 148p$	89p
	241p
Profit and Oncost 20%	48p
	289p
Rate per m	£2·89

Example 18

250 x 100 mm precast-concrete sill, sunk weathered, throated, finished fair on three surfaces, girth 275 mm, reinforced with two 6 mm mild steel rods.

Cost of sill per m £2·75	275p
Allow for waste $2\frac{1}{2}\%$	7p
From Example 16 a comparative rate for unloading, hoisting, and setting etc. may be deduced.	
Total cost for setting a 100 x 250 mm lintel = 148p	
∴ Cost for a setting 250 x 100 mm sill =	
$\frac{250 \times 100}{100 \times 250} \times 148p$	148p
	430p
Profit and Oncost 20%	86p
	516p
Rate per m	£5·16

Example 19

50 mm precast concrete paving slab in size 600 x 600 mm laid with risband joints on 50 mm bed of sand and jointed and pointed with cement mortar (1:3).

Material	
1 m² of paving slabs cost	181p
Unload and stack 50 per hour	11p
$\frac{1}{16}$ m³ sand (to allow for sand being compressed) @ 319p per m³	20p
Allow for cement mortar for jointing, pointing, say	5p
	217p
Waste 5%	11p
C/fd	228p

B/fd	228p
Labour	
Two tradesman to one labourer =	
2/220p + 190p = 630p	
One tradesman lays 1·67 m² per hour @ 315p:	
∴ 1 m² @ 315p ÷ 1·67	189p
	417p
Profit and Oncost 20%	83p
	500p
Rate per m²	£5·00

Example 20

Concrete bed exceeding 100 mm but not exceeding 150 mm thick to comply with Concrete Grade 20, 20 mm aggregate, medium workability, Zone 1%, in accordance with B.S. 5328: 1976 and C.P. 110: Part 1: 1972.

The materials required by weight to produce approximately 1 m³ of fully compacted concrete in accordance with the B.S. and C.P. are:

Cement 320 kg @ 2,552p per tonne (*See* Example 3)	817p
Sand 720 kg @ 210p per tonne	151p
Aggregate, 1,080 kg @ 250p per tonne	270p
Materials cost for 1 m³ of compacted concrete	1,238p
Mixing, using 7/5 mixer (*See* Example 4)	302p
Transporting and placing, 5 hours @ 190p	950p
	2,490p
Profit and Oncost 20%	498p
	2,988p
Rate per m³	£29·88

The difference in cost from Example 4 is 7p

7 Brickwork

COMMON BRICKWORK

BS 3921: Part 2: 1969 is the British Standard Specification for Bricks and Blocks of Fired Brickearth, Clay or Shale.

Whilst specifying the required standards, it also defines the dimensions of bricks in Table 1, shown below:

Table 1 Standard Formats (Bricks)

Designation	Actual dimensions (mm) Length	Width	Height
225 x 112·5 x 75	215	102·5	65

Table 3 of this BS defines dimensional tolerances allowable as follows:

Table 3 Dimensional Tolerances (Bricks)

Specified dimension (Table 1) mm	Overall measurement of 24 Bricks mm
65	1560 { + 60 / − 30
102·5	2460 ± 45
215	5160 ± 75

As can be seen from the Table, the measurements without tolerances are for bricks set dry, without mortar beds or joints.

Bricks which conform to BS 3921 are generally referred to as common bricks.

In computing rates for brickwork the estimator must know the following:

(1) The quantity of bricks required per square metre for the required thickness of walling; also the waste that may be incurred.
(2) The amount of time necessary for a bricklayer and labourer to build the required quantity.
(3) The quantity of mortar necessary to build the required amount of bricks.

It is proposed that bricks four courses in height will equal 300 mm and that four bricks in length will equal 900 mm.

Thus the metric brick is 215 x 102·5 x 65 mm, which, with a 10 mm joint, produces a working or format size of 225 x 112·5 x 75 mm, which is very close to the old imperial working size of 9 x $4\frac{1}{2}$ x 3″.

To calculate the No. of Bricks with 10 mm joint per m^2

$$225 \times 75 \text{ mm} = 16{,}875 \text{ mm}^2$$
$$1{,}000{,}000 \text{ mm}^2 = 1 \text{ m}^2$$

$$\therefore \text{No. of bricks per m}^2 = \frac{1{,}000{,}000}{16{,}875} = 59{\cdot}26$$

This for practical purposes = 59 bricks.

No. of 65 mm bricks per m^2

Half-brick wall	59
One-brick wall	118
One and half brick wall	177
Reduced brickwork	118

PRODUCTIVITY

There is considerable diversity of opinion as to how many common bricks can or should be laid by a tradesman per hour. This varies from firm to firm and the output used by the estimator is a product of years of experience coupled with good ‘feed-back’ information from the various sites and based on local Trade Union agreement. A reasonable average is:

Type of wall	Average output per tradesman per hour
Half-brick wall	55
One-brick wall	55
One and half brick wall	60
Reduced brickwork	65
Brickwork in projections	50
Brickwork in backing to masonry	55
Facing brickwork	40
Engineering brickwork in manholes	50
Engineering brick as facing brick	40
Labours	
Rough arch (waste 12 bricks per m^2)	30
Key or flush pointing as work proceeds	3 m^2
Rake out and point at a later date	2 m^2
Form cavity in hollow walls including placing wall ties	12 m^2
Rough cutting (waste 11 bricks per m^2)	1 m^2

Labours on Facing Brick

	Tradesman hours per lin.m	Brick waste per lin.m
Fair cutting at vertical abutments	0·33	3·50
Fair cutting at soffits	0·50	2·00
Fair raking or splay cutting	0·50	2·00
Key or flush pointing as work proceeds	1·70 m^2	
Rake out and point at a later date	1·00 m^2	

MORTAR

Brickwork is generally built either with cement mortar or cement-lime mortar, and on occasion with lime mortar. For the purpose of this publication we will consider cement mortar, and cement-lime mortar. Cement may be delivered loose in bulk, or may be delivered in paper sacks. If delivered in bulk it must be stored on site in silos and no cost is incurred for unloading, but the cost of the use, or hire, of the silo must be charged in the Preliminaries. If delivered in paper sacks the cost of unloading must be dealt with in computing the cost of mortar. Lime is delivered to site in paper sacks and a reasonable allowance for unloading and stacking in a shed for both lime and cement is 1 hour of a labourer's time per $1\frac{1}{4}$ tonnes.

When mixing mortar the introduction of water to the mix causes a decrease in bulk which is generally referred to as shrinkage. The shrinkage for mortar composed of cement and sand is one fifth or 20%; and for cement-lime-sand mortar, one quarter or 25%.

A reasonable time for hand-mixing one cube metre of mortar is 5 hours of a labourer's time; machine mixing is shown in subsequent examples.

In the following examples it will be seen that the method used in computing rates is based on the unit of 1,000 bricks, which unit is subsequently reduced to square metres of the required thickness of brickwork.

The reason is to eliminate unnecessary calculations. Bricks are purchased by the 1,000, cement and lime by the tonne; and labour output constants are normally based on the number of bricks built per hour. To reduce each of these units to the equivalent of a square metre involves detailed calculations for each one. By using the unit of 1,000 bricks, one has to hand the cost without calculation, and the cost of unloading (if necessary) is added. The cost of mortar is normally assessed by the cubic metre, and the quantity required to build 1,000 bricks, expressed in cubic metres, is added. The unit of 1,000 bricks is divided by the number of bricks the estimator considers the bricklayer will build per hour, which gives the number of hours required to build 1,000 bricks.

For the purpose of this publication the writers are using the constant of two bricklayers being attended by one labourer, exclusive of the *mixing* of mortar and the provision of scaffolding, which involve additional labour service. The writers are aware that the amount, or ratio, of labour required to service bricklayers may vary, dependent on the height and number of storeys in the building and also the location of the site, and that the opinions of estimators are at variance on this question; but they are satisfied that the assumptions made for this publication are fair and reasonable.

Scaffolding is not taken into account at this stage as it should normally be dealt with in the Preliminaries. That is, external scaffolding is assessed from the description of the building detailed in the Preliminaries and internal scaffolding (normally loose battens and trestles) is assessed from the area of partitions.

The quantities of mortar shown for building 1,000 bricks include an allowance of about 10% for waste, and although the machine mixed mortar is shown to be a hired machine the cost of machine may also be shown on the basis of contractor-owned mortar mixer; and the average hourly cost of machine may be assessed in the same way as that shown in the chapter on concrete.

MORTAR EXAMPLES

Example 1

Cement mortar (1:3). Because of the nature of the job it has been taken that the mortar will be hand-mixed.

Material			
Portland cement d/d site per tonne (1,000 kg)	2,400p		
Labour unloading $1\frac{1}{4}$ tonnes per hour @ 190p	152p		
	2,552p		
1 m³ cement = 1·44 tonnes @ 2,552p per tonne			3,675p
3 m³ sand = 4·56 tonnes @ 210p per tonne			958p
			4,633p
Shrinkage is 20%, ∴ add 25%			1,158p
4 m³ =			5,791p
∴ 1 m³ of material costs		C/fd	1,448p

	B/fd	1,448p
Labour		
5 hours of labourer mixing @ 190p		950p
∴ Cost of 1 m^3 without profit		2,398p

Example 2

Machine-mixed mortar (1:3).

Material As before		1,448p
Labour		
5/3½ (0·140/0·100 m^3) mixer hire rate is £9·50 per week	£9·50	
1 labourer in charge 190p plus 1p x 40 =	£76·40	
1 labourer in attendance 190p x 40 =	£76·00	
Cost per week	£161·90	
Machine is used 37 hours, allowing 3 hours standing time per week		
∴ Cost per hour = £161·90 ÷ 37 =	438 p	
Cleaning and maintaining ½ hour per day of 8-hour day $\frac{1}{16}$ of 191p =	12 p	
Fuel 1·14 litres petrol @ 18p	20½p	
Oil and grease (not included in hire rates), say	7½p	
	478p	
Output		
1·15 m^3 per hour for mortar (1·00 m^3 for concrete)		
∴ Cost of mixer per m^3 = 478p ÷ 1·15		416p
∴ Cost of 1 m^3 without profit		1,864p

Alternative to Example 2

Cost of mixer per week with no attendance = £85·90 (£161·90 less £76·00).

Cost per week = £85·90 ÷ 37	232p
Cleaning, fuel, oil, and grease, etc.	40p
	272p

Output
0·59 m^3 per hour with one man working
∴ Cost of mixer per m^3 using only 1 man

= 272p ÷ 0·59 =	461p
Material As before	1,448p
∴ Cost of 1 m^3 without profit	1,909p

Example 3

Cement-lime mortar (1:1:6) hand-mixed

Lime d/d site per tonne	2,800p per tonne
Unload $1\frac{1}{4}$ tonnes per hour @ 190p	152p
	2,952p

1 m^3 cement 1·44 tonnes @ 2,552p	3,675p
1 m^3 lime 0·60 tonnes @ 2,952p	1,771p
6 m^3 sand 9·12 tonnes @ 210p	1,915p
	7,361p
Shrinkage is 25%, ∴ add $\frac{1}{3}$	2,454p
8 m^3 costs	9,815p
∴ 1 m^3 costs	1,227p
Labourer mixing 5 hours @ 190p	950p
∴ Cost of 1 m^3 without profit	2,177p

Example 4

Cement-lime mortar (1:1:6) machine-mixed

Material	
Example 3	1,227p
Labour	
Example 2	416p
Cost of 1 m^3 without profit	1,643p

BRICKWORK EXAMPLES

Example 5

Common brickwork half-brick thick in cement mortar (1:3).

65 mm common bricks d/d site per 1,000		3,100p
Allow for waste 5%		155p
		3,255p
Labour		
Based on 2 tradesmen and 1 labourer		
∴ 2 tradesmen for 1 hour @ 220p	440p	
1 labourer for 1 hour @ 190p	190p	
	630p	
Output per hour = 55 per tradesman		
∴ 1,000 costs $\frac{1,000}{110} \times 630p$		5,727p
Mortar (1:3) *machine mixed*		
Example 2 alternative		
1 m^3 costs 1,909p		
0·53 m^3 of mortar builds 1,000 bricks in half-brick walls		
∴ Cost of mortar is 1,909p x 0·53 =		1,012p
Cost of 1,000 bricks		9,994p

No. of bricks per m^2 = 59

$\therefore$ Cost of 1 $m^2 = \frac{9{,}994}{1{,}000} \times 59$	590p
Profit and Oncost 20%	118p
	708p
Rate per m^2	£7·08

Example 6

Brickwork 1 brick thick built in (1:3) cement mortar (hand-mixed).

65 mm common bricks as before	3,255p
Labour	
Squad 2 tradesmen and 1 labourer	
Output per hour = 55 per tradesman	
$\therefore$ 1,000 costs $\frac{1{,}000}{110} \times 630$p	5,727p
Hand-mixed Mortar	
As before	
0·60 m^3 of mortar builds 1,000 bricks in 1 brick wall	
$\therefore$ Cost of mortar = 2,398p x 0·60	1,439p
	10,421p

No. of bricks per m^2 = 118

$\therefore$ Cost of 1 $m^2 = \frac{10{,}421}{1{,}000} \times 118$	1,230p
Profit and Oncost 20%	246p
	1,476p
Rate per m^2	£14·76

Example 7

Brickwork half-brick thick in cement lime mortar (1:1:6)

Bricks	
As before	3,255p
Labour	
As before	5,727p
Mortar 1 m^3 costs 1,620p	
∴ 0·53 per 1,000 costs 1,643p x 0·53	871p
	9,853p

No. of bricks per m^2 = 59

∴ Cost of 1 m^2 $= \frac{9{,}853}{1{,}000} \times 59$	581p
Profit and Oncost 20%	116p
	697p
Rate per m^2	£6·97

Example 8

Brickwork one and a half bricks thick in cement mortar (1:3).

Bricks	
As before	3,255p
Labour	
2 tradesmen and 1 labourer	
Output per hour = 60 per tradesman	
∴ 1,000 costs $= \frac{1{,}000}{120} \times 630$p	5,250p
Mortar from Example 2: m^3 costs 1,841p	
∴ Cost of mortar per 1,000 = 1,864 x 0·60	1,118p
	9,623p

No. of bricks per m² = 177

∴ Cost of 1 m²	$= \frac{9,623}{1,000} \times 177$	1,703p
	Profit and Oncost 20%	341p
		2,044p
	Rate per m²	£20·44

Example 9

Common brickwork reduced to one brick thick (1:3).

Bricks	
As before	3,255p
Labour	
2 tradesmen and 1 labourer	
Output per hour 65 per tradesman	
∴ 1,000 costs $\frac{1,000}{130} \times 630p$	4,846p
Mortar as Example 8	1,118p
	9,219p

No. of bricks per m² = 118

∴ Cost of 1 m²	$= \frac{9,219}{1,000} \times 118$	1,088p
	Profit and Oncost 20%	218p
		1,306p
	Rate per m²	£13·06

Example 10

Brickwork in projections of chimney breasts reduced to one brick thick (1:3)

Bricks		
As before	C/fd	3,255p

Labour

2 tradesmen and 1 labourer	B/fd	3,255p
Output per hour 50 per tradesman		
$\therefore$ 1,000 costs $\frac{1,000}{100} \times 630$p		6,300p
Mortar as Example 8		1,118p
		10,673p
No. of bricks per m^2 = 118		
$\therefore$ Cost of 1 m^2 = $\frac{10,673}{1,000} \times 118$		1,259p
Profit and Oncost 20%		252p
		1,511p
Rate per m^2		£15·11

Example 11

Form cavity 75 mm wide in hollow walls including 4 galvanized wall ties per m^2

Wall ties cost £48·30 per 100 kg (700 wall ties)		
$\therefore$ 1 m^2 costs $\frac{4,830}{700} \times 4$		$27\frac{1}{2}$p
1 tradesman setting and clearing cavity 5 minutes @ 220p		$18\frac{1}{2}$p
	C/fd	46 p
Profit and Oncost 20%		9p
Rate per m^2		55p

Example 12(a)

Rough cutting: on common brickwork

Rough cutting; tradesman 1 hour @ 220p		220p
Waste on bricks 11 bricks per m^2 @ 3,255p per 1,000		36p
	C/fd	256p

	B/fd	256p
Profit and Oncost 20%		51p
		307p
Rate per m^2		£3·07

Example 12(b)

Brickwork in projections of plinths set projecting 56 mm from face of wall and 600 mm wide (1:3).

Note: Rough cutting is deemed to be included

Bricks

As before	3,255p

Labour

2 tradesmen and one labourer

Output per hour = 50 per tradesman

$\therefore$ 1,000 costs $\frac{1,000}{100} \times 630p$	6,300p
Mortar as Example 8	1,118p
	10,673p

No. of bricks per m^2 one brick thick = 118	
$\therefore$ Cost of 1 m^2 = $\frac{10,673}{1,000} \times 118$	12,59p

Cost of 1 m^2 $\frac{1}{4}$ brick thick = 1,259p ÷ 4	315p
Rough cutting: tradesman 1 hour @ 220p	220p
Waste on bricks 11 per m^2 @ 3,255p per 1,000	36p
Cost of 1 m^2	571p

$\therefore$ Cost of 1 m 600 mm wide = $\frac{571p}{1,000} \times 600$	343p
Profit and Oncost 20%	69p
	412p
Rate per m (600 mm wide)	£4·12

Example 13

Close 75 mm wide cavity of hollow wall with brickwork half brick thick

Allow 5 bricks per m, which includes for waste.

Cost of 5 bricks @ 3,255p per 1,000 is $\frac{3,255}{1} \times \frac{5}{1,000}$		16p
Mortar 0·003 m³ @ 1,643p per m³ (*see* Example 4)		5p
		21p
Bricklayer roughcutting 6 minutes @ 220p per hour	22p	
Bricklayer building 6 minutes @ 220p per hour	22p	
Labourer 3 minutes @ 190p per hour	10p	54p
		75p
Profit and Oncost 20%		15p
Rate per m		90p

Example 14

Extra over common brickwork for rough segmental arch two rings high and one brick wide on soffit including rough cutting on arch and wall.

Note: Width of brick built in arch is 65 + 10 mm = 75 mm on face.

Number of bricks per lin.m of arch = $\frac{1,000}{75} \times 2$ rings $= 26\frac{2}{3}$; say 27.

Output on one brick wall–55 bricks per hour of bricklayer and half a labourer

Output on arch–30 bricks per hour of bricklayer and half a labourer

Cost of building 27 bricks in arch $= \frac{27}{30} \times 315$p	C/fd	284p

	B/fd	284p
Deduct Cost of building 27 bricks in wall = $\frac{27}{55} \times 315$p		155p
Extra cost of building per m		129p
Rough cutting on wall: 14 minutes of bricklayer @ 220p per hour		51p
Waste on bricks 20%, i.e. 2 bricks per lin.m @ 3,100p per 1,000		6p
		186p
Profit and Oncost 20%		37p
		223p
Rate per m		£2·23

BRICK FACEWORK

For walls built in common brickwork and faced and pointed with common brick selected from stacks, no extra materials are required. The only additional cost is extra labour selecting bricks and pointing as the work proceeds. The following example illustrates this:

Example

Extra for fair face and pointing to common brickwork as the work proceeds.

Pointing brickwork 3 m² per hour ∴ 1 tradesman 20 minutes @ 220p	73p
Labourer selecting; 5 minutes @ 190p	16p
	89p
Profit and Oncost 20%	18p
	107p
1 m² costs	£1·07

EXTRA OVER WORK

Where walls are faced with facing bricks which are different from those built in the body of the work, the bricks in the body of the work normally being built of common bricks, the facework is measured by the square metre *as extra over* the brickwork on which it occurs.

Extra over means the additional difference in cost between the cost of common bricks built in mortar and facing bricks built in the same mortar plus pointing.

Facing bricks are normally bonded into the common brick backing and the type of bond determines the amount of facing brick required per square metre of walling. This is illustrated as follows:

$$\text{One square metre} = 1{,}000 \times 1{,}000 \text{ mm} = 1{,}000{,}000 \text{ mm}^2$$

One brick is 215 mm long and 65 mm high.
One brick with 10 mm bed and joint = 225 mm long and 75 mm high.
∴ the number of bricks per m^2 built stretcher bond

$$\frac{1{,}000{,}000}{225 \times 75} = 59{\cdot}26.$$

This for practical purposes = 59 bricks.

Scottish or Garden Wall bond is three stretcher courses to one header course, thus:

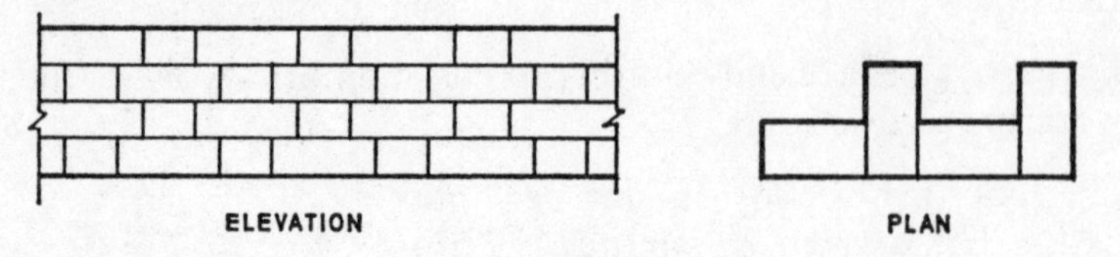

Facing brick displacing common brick is 25% more = 59·26 + 25%.

This for practical purposes = 74 bricks.

Flemish bond is illustrated thus:

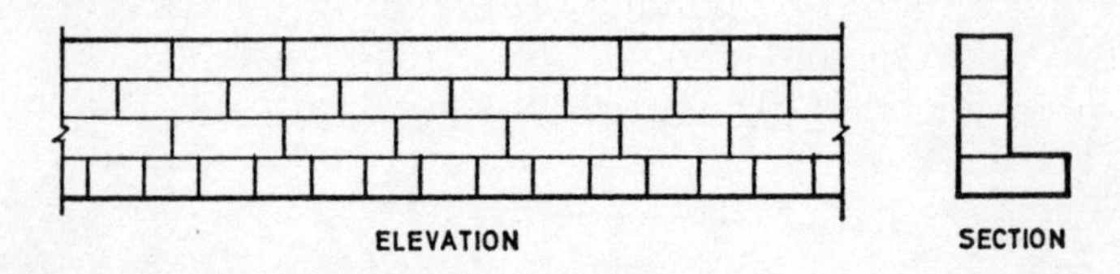

Facing brick displacing common brick is $33\frac{1}{3}\%$ more $= 59{\cdot}26 + 33\frac{1}{3}\%$.

This for practical purposes = 79 bricks.

English bond is illustrated thus:

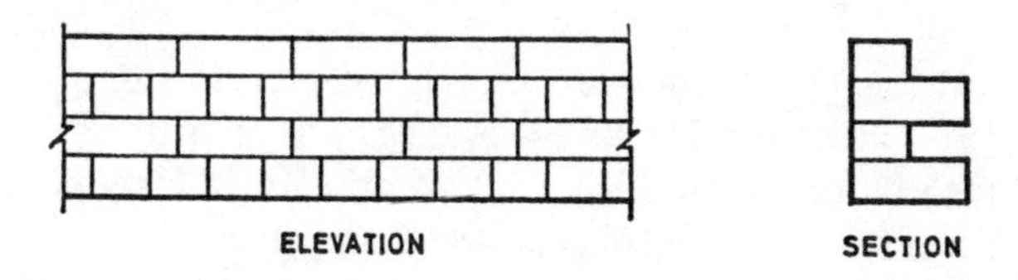

Facing brick displacing common brick is 50% more $= 59{\cdot}26 + 50\%$.

This for practical purposes = 89 bricks.

When computing a rate for facing bricks some estimators assume that the labour for building common bricks is the same as for building facing bricks. In making this assumption the amount of facing bricks required for a square metre of facework is determined, e.g. 89 for English bond. The difference in cost between 89 common and 89 facing bricks is then assessed and the extra cost of labour pointing brickwork is added, and a percentage for profit is added to the total, which gives the extra over rate for facing brick. This method is not recommended, as a bricklayer normally takes more time to build facing bricks than common bricks; therefore it will not give an accurate assessment of the extra over price of facework.

The method suggested is to take the difference in cost of 1,000 common bricks and 1,000 facing bricks delivered to site and the difference in cost of building, and add these figures, which gives the extra over cost for a 1,000 facing bricks built. The extra over cost for one square metre is then determined by proportioning the requisite square metre number to the extra cost of 1,000 bricks.

There are two methods of computing the rate for brick facework:

(1) Is to find the difference in cost between the facing bricks and commons, also the difference in cost of labour building the bricks. The costs are then totalled and profit is added. In this method the cost of mortar, other than for coloured mortar for pointing, is not taken into account as the

mortar for building the facing bricks is normally the same as used for the common brickwork in the body of the work.

(2) Is to build up a full value rate for facing bricks and to deduct the cost of the same number of common bricks built and then add profit, which gives the extra over rate for facing brick.

Method 1 is the method which will be used in this publication, although Method 2 is shown as an alternative. It should also be noted that by using Method 1 an Extra Over rate can be built up, say for examination purposes, for facing bricks when being given only the costs of the facing and common bricks, whereas, if using Method 2, one must know the rate for common bricks or must build up a rate for common bricks before being in a position to determine the extra over for facing bricks.

Example 15

METHOD 1

Extra over common brickwork for facing with Golden Brown Sandfaced rustic bricks (Prime Cost d/d site £60·00 per 1,000) built Flemish bond and key-pointed as work proceeds.

Cost of facing bricks per 1,000 d/d site	6,000p		
Unload and stack; labourer 2 hours @ 190p per hour	380p		
	6,380p		
Allow waste 5%	319p		
	6,699p		
Deduct cost of 1,000 common bricks d/d site (including waste)	3,255p		
∴ Extra cost of 1,000 facings	3,444p		3,444p
		C/fd	3,444p

	B/fd	3,444p
Labour cost of building 1,000 facings 2 bricklayers and 1 labourer output per bricklayer hour = 40		
$\therefore \frac{1,000}{80} \times 630p$	7,875p	
Deduct		
Labour cost of building 1,000 commons 2 bricklayers and 1 labourer output per bricklayer hour = 55		
$\therefore \frac{1,000}{110} \times 630p$	5,727p	
Extra cost of building 1,000 facings	2,148p	2,148p
Total extra cost of building 1,000 facings		5,592p

Number of bricks per m^2 (Flemish Bond) = 79
$\therefore$ extra cost of 1 m^2 of facings =

$\frac{79}{1,000} \times 5,592p$	442p
Bricklayer points 1·70 m^2 per hour, $\therefore \frac{220p}{1 \cdot 70}$	129p
	571p
Profit and Oncost 20%	114p
	685p
Extra over rate per m^2	£6·85

METHOD 2

Extra over common brickwork for facing with Golden Brown Sandfaced rustic bricks (Prime Cost d/d site £60·00 per 1,000) built Flemish bond and key-pointed as work proceeds.

Cost of facing bricks per 1,000 d/d site	6,000p
Unload and stack; labourer 2 hours @ 190p	380p
	6,380p
Allow waste 5%	319p
Output based on 40 bricks per hour	
∴ Cost of building 1,000 facing bricks = $\frac{1,000}{80} \times 630p$	7,875p
Mortar for building 0·60 m^3 @ 2,398p (*see* Example 1)	1,439p
	16,013p

Number of facing bricks per m^2 (Flemish bond) = 79	
Cost of 79 facing bricks = $\frac{16,013}{1,000} \times 79$	1,265p
Bricklayer points 1·70 m^2 per hour, ∴ $\frac{220}{1 \cdot 70}$	129p
	1,394p
Deduct	
Equivalent amount of common bricks displaced (*see* Example 6) ∴ 79 @ 1,230p for 118	823p
	571p
Profit and Oncost 20%	114p
	685p
Extra over rate per m^2	£6·85

Example 16

Extra over common brickwork for facing with multi-coloured facing bricks (Prime Cost d/d site £60·00 per 1,000) built English bond, raked out and key-pointed at a later date with tinted cement mortar composed of one part Snowcrete to three parts silver sand.

Cost of facing bricks per 1,000 d/d site		6,000p
Unload and stack; labourer @ 190p per hour		380p
		6,380p
Allow waste 5%		319p
		6,699p
Deduct cost of 1,000 common bricks d/d site (including waste)		3,255p
Extra cost of 1,000 facing bricks		3,444p

Labour
Based on 2 tradesmen to 1 labourer
@ 630p per hour
Output per hour 40 per tradesman
Labour cost of building 1,000

facing bricks = $\frac{1,000}{80} \times 630p$	7,875p	
Deduct		
Labour cost of building 1,000 common bricks $\frac{1,000}{110} \times 630p$	5,727p	
Extra cost of building 1,000 facing bricks	2,148p	2,148p
Total extra cost of 1,000 facing bricks		5,592p

Number of facings per m²
(English bond) = 89
∴ Extra cost of 1 m² of facing bricks =

$\frac{89}{1,000} \times 5,592p$		498p

Mortar for pointing

Snowcrete per tonne d/d site in small loads	6,200p	
C/fd	6,200p	498p

B/fd	6,200p	498p
Unload and stack 1·25 tonnes per hour labourer @ 190p	152p	
	6,352p	
1 m^3 Snowcrete = 1·44 tonnes @ 6,352p	9,147p	
3 m^3 silver sand = 4·56 tonnes @ 700p	3,192p	
	12,339p	
Shrinkage is 20%, ∴ add 25%	3,085p	
Cost of 4 m^3	15,424p	
∴ 1 m^3 of material costs	3,856p	
Mixing; labourer 5 hours @ 190p	950p	
	4,806p	
Mortar for pointing: requires 0·033 m^3 per 1,000 bricks @ 4,806p per m^3;		
∴ cost per m^2 = $0{\cdot}033 \times \frac{59}{1{,}000} \times 4{,}806$p		9p
Labour pointing 1·00 m^2 per hour of bricklayer @ 220p		220p
		727p
Profit and Oncost 20%		145p
		872p
Extra over rate per m^2		£8·72

Example 17

Extra over common brickwork for facing with multi-coloured facing bricks (Prime Cost d/d site £60·00 per 1,000) built English bond to sunk panels set back 56 mm from face of wall, raked out and key-pointed at a later date with tinted mortar cement composed of one part Snowcrete to three parts silver sand.

(*See* Example 16 for cost of facing brickwork.)

Extra cost of facing bricks (without profit) per m² from Example 16	727p
Setting brickwork back 56 mm; 15 minutes bricklayer @ 220p per hour	55p
Rough cutting on common brickwork; 1 hour bricklayer @ 220p per hour	220p
Waste on brickwork nil	
	1,002p
Profit and Oncost 20%	200p
	1,202p
Extra over rate per m²	£12·02

Note: (1) Waste is nil in the above case as it is included in the measurement of common brickwork. *See* S.M.M.

(2) Facework in panels is measured extra over brickwork on which it occurs. *See* S.M.M.

Example 18

Extra over common brickwork for facing with multi-coloured facing bricks (Prime Cost d/d site £60·00 per 1,000) built English bond to panels set projecting 56 mm from face of wall, raked out and key-pointed at a later date with tinted cement mortar composed of one part of Snowcrete to three parts silver sand.

(*See* Example 16 for cost of facing brickwork.)

Extra cost of facing bricks (without profit) per m² from Example 16		727p
Cost of common brickwork from Example 9, assuming the wall is reduced brickwork = 414p (without profit) ∴ Cost is 1,088p ÷ 4 to obtain a quarter brick =	272p	
C/fd	272p	727p

B/fd	272p	727p
Waste on brickwork, 20% = $\frac{3{,}100p}{1{,}000} \times 59 \times \frac{1}{5}$	37p	309p
Setting brickwork forward 56 mm: 15 minutes bricklayer @ 220p per hour		55p
Rough cutting on *common brickwork*: 1 hour bricklayer @ 220p per hour		220p
		1,311p
Profit and Oncost 20%		262p
		1,573p
Extra over rate per m²		£15·73

Note: Reduced brickwork is one brick thick as shown at Example 9, therefore the brickwork is divided by 2 to bring to half-brick thick and divided again by 2 to give the required 56 mm thickness of projection. It should be noted that the extra thickness described in accordance with G.15.12 is made up of common brickwork.

Example 19

Extra over common brickwork for facing to semi-circular arch one brick wide on face and half brick wide on soffit with purpose-made facing brick voussoirs (Prime Cost d/d site £75·00 per 1,000) built in cement mortar (1:3) and key-pointed with coloured cement mortar at a later date.

Cost of purpose-made facing bricks per 1,000 d/d site	7,500p
Unload and stack: labourer 2 hours @ 190p	380p
	7,880p
Allow waste 5%	394p
C/fd	8,274p

	B/fd	8,274p
Output based on 20 bricks per hour per tradesman		
∴ Cost of building 1,000 voussoirs		
$= \frac{1,000}{40} \times 630p$		15,750p
Mortar 0·60 m³ @ 1,864p (*see* Example 2)		1,118p
Cost of 1,000 voussoirs in semi-circular arch		25,142p

Note: The bricks are measured mean length on face and the voussoir is thus:

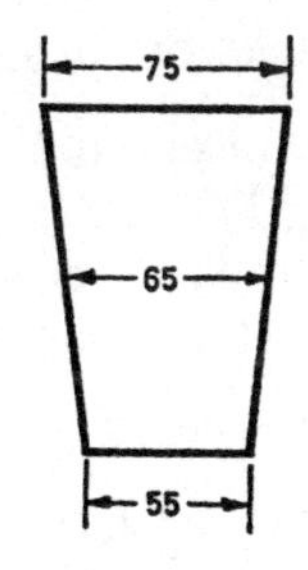

The brick and bed are 75 mm wide on extrados and about 55 mm wide on intrados, therefore 65 mean width, which gives 1,000 ÷ 65 = 15·4 bricks per lin.m of arch.

∴ Per lin.m = $\frac{25,142p}{1,000} \times 15{\cdot}4$		387p
Mortar for pointing, say		3p
Bricklayer pointing based on 1·00 m² per hour @ 220p per hour		
Girth of arch = 338 mm 1·00 m long = 0·40 m²		
∴ 220p x 0·40 ÷ 1·00		88p
		478p
From Example 6, cost of building 118 common bricks	1,230p	
∴ Cost of building 13⅓ common bricks	139p	C/fd 478p

B/fd	478p
As there are $13\frac{1}{3}$ common bricks in one linear metre, one brick high, *Deduct* cost of $13\frac{1}{3}$ common bricks (without profit)	139p
	339p
Profit and Oncost 20%	68p
	407p
Extra over rate per m	£4·07

Example 20

Fair cutting on facing brickwork at vertical abutments.

Bricklayer cuts 3·00 m per hour; ∴ 1 lin.m costs 220p ÷ 3	73p
Waste–Allow 3·50 bricks @ 6,699p per 1,000 (*see* Example 15)	24p
	97p
Profit and Oncost 20%	19p
Rate per m	£1·16

Example 21

Facework to margins including pointing as work proceeds.

Note: The labour involved is plumbing angle and pointing.

Bricklayer plumbs 9 lin.m per hour ∴ Plumbing cost per lin.m 220p ÷ 9 =	24p
Pointing as work proceeds is 1·70 m^2 per hour; assume 1·00 m^2 per hour for narrow widths:	
∴ pointing 1 lin.m 113 mm wide $220p \times \frac{113}{1{,}000} \times 1$	25p
C/fd	49p

	B/fd	49p
Profit and Oncost 20%		10p
Rate per m		59p

GLAZED BRICKS

Glazed bricks may be obtained in various colours, and are purchased glazed on one face, one end, or one face and one end.

There are also various types of bullnosed bricks.

The stretchers (glazed one face), and the headers (glazed one end) are a different price per 1,000, therefore adjustment must be made for this.

The production rate for glazed bricks is about three quarters that of ordinary facing bricks. The output taken here is 30 per hour of bricklayer.

On considering bond the actual amount of bricks for each type of bond is thus:

English bond	*Flemish bond*	*Garden wall bond*
	Per square metre	
$59\frac{1}{3}$ Headers	$39\frac{1}{3}$ Headers	$29\frac{1}{2}$ Headers
$29\frac{2}{3}$ Stretchers	$39\frac{1}{3}$ Stretchers	$44\frac{1}{4}$ Stretchers
89 Net total	$78\frac{2}{3}$ Net total	$73\frac{3}{4}$ Net total
	Per 1,000	
$666\frac{2}{3}$ Headers	500 Headers	400 Headers
$333\frac{1}{3}$ Stretchers	500 Stretchers	600 Stretchers
1,000	1,000	1,000

Example 22

Extra over common brickwork for facing with white glazed facing bricks built English bond and key-pointed as the work proceeds. (Header bricks cost £280·00 and stretcher bricks cost £285·00 per 1,000 d/d site.)

Cost of facing bricks per 1,000

Header bricks per 1,000 d/d site cost £280·00 ∴ £280·00 x $\frac{2}{3}$ =		18,666p
Stretcher bricks per 1,000 d/d site cost £285·00 ∴ £285·00 x $\frac{1}{3}$ =		9,500p
		28,166p
Unload and stack: 2 hours @ 190p		380p
Average cost per 1,000		28,546p
Allow waste 5%		1,427p
		29,973p
Deduct cost of common bricks		3,255p
		26,718p
Extra cost of 1,000 glazed bricks built English bond		26,718p
Labour: based on 2 bricklayers and 1 labourer output based on 30 bricks per hour		
∴ Cost of building 1,000 facings = $\frac{1{,}000}{60}$ x 630p	10,500p	
Deduct Cost of building 1,000 commons = $\frac{1{,}000}{110}$ x 630p	5,727p	
Extra cost of building 1,000 facings	4,773p	4,773p
Total extra cost of 1,000 facings		31,491p
Number of facings per m^2 (English bond) = 89		
∴ Extra cost of 1 m^2 of facings = $\frac{89}{1{,}000}$ x 31,491p		2,803p
Bricklayer points 1·70 m^2 per hour ∴ 220p ÷ 1·70		129p
	C/fd	129p

	B/fd	2,932p
Profit and Oncost 20%		586p
		3,518p
Extra over rate per m²		£35·18

Example 23

Form bullnosed angle on white-glazed facing bricks to 50 mm radius

Note: Extra cost of bullnosed bricks over glazed bricks is £70·00 per thousand.

∴ Extra cost of $13\frac{1}{3}$ white-glazed bullnosed bricks = $\frac{7{,}000p}{1{,}000} \times 13\frac{1}{3}$	93p
Allow Waste 5%	5p
Bricklayer plumbs 4·50 lin.m per hour ∴ per lin.m 220p ÷ 4·50 =	49p
	147p
Profit and Oncost 20%	29p
Rate per m	£1·76

Note: $13\frac{1}{3}$ bricks built in courses 75 mm high represent one linear metre of angle. The extra labour setting is included in the rate for plumbing.

BRICKWORK ENTIRELY OF FACING BRICK

Example 24

Half brick wall in skins of hollow walls built entirely of 65 mm multi-coloured stock facing bricks (Prime Cost £65·00 per 1,000 d/d site), built stretcher bond in cement-lime mortar 1:1:6 and weather pointed one face as the work proceeds.

Cost of facing bricks per 1,000 d/d site	6,500p
Unload and stack; labourer 2 hours @ 190p per hour	380p
	6,880p
Allow waste 5%	344p
	7,224p
Labour based on 2 bricklayers to 1 labourer; based on output of 40 bricks per hour ∴ Cost of building 1,000 facings = $\frac{1{,}000}{80} \times 630p$	7,875p
Mortar: 0·53 m^3 builds 1,000 bricks, ∴ 1,643p (Example 4) x 0·53	871p
Cost of 1,000 facing bricks built	15,970p
Number of bricks per m^2 of half brick wall = 59 ∴ $\frac{15{,}970p}{1{,}000} \times 59$	942p
Bricklayer points 1·70 m^2 per hour, ∴ 1 m^2 = 220p ÷ 1·70	129p
	1,071p
Profit and Oncost 20%	214p
	1,285p
Full value rate per m^2	£12·85

Note: Brickwork built entirely of facing brick applies mainly to half brick and one brick thick walls and is measured full value and not extra over common brickwork; *see* S.M.M.

Example 25

One-brick wall built entirely of 65 mm facing bricks (Prime Cost d/d site £60·00 per 1,000) in Flemish bond in cement-lime mortar (1:1:6) and struck pointed on both faces as the work proceeds.

Note: In a one-brick wall built entirely of 65 mm facing bricks the quantity of bricks is always the same, regardless of bond, i.e. 118 No.

Cost of facing bricks per 1,000 d/d site		6,000p
Unload and stack; labourer 2 hours @ 190p per hour		380p
		6,380p
Allow waste 5%		319p
		6,699p
Labour based on 2 bricklayers and 1 labourer; based on output of 40 brick per hour		
∴ Cost of building 1,000 facings = $\frac{1{,}000}{80} \times 630$p		7,875p
Mortar (Example 4): $0{\cdot}60\ m^3$ builds 1,000 bricks in one-brick wall		
∴ 1,643p x 0·60		986p
Cost of 1,000 facings built		15,560p
Number of bricks per m^2 of one brick wall	118	
∴ $\frac{15{,}560p}{1{,}000} \times 118$		1,836p
Bricklayer points $1{\cdot}70\ m^2$ per hour		
∴ $1\ m^2 = (220p \div 1{\cdot}70) \times 2$ sides		259p
		2,095p
Profit and Oncost 20%		419p
		2,514p
Full value rate per m^2		£25·14

Example 26

Flush brick-on-edge cope of 65 mm facing bricks (Prime Cost d/d £60·00 per 1,000) 225 mm broad and 113 mm high built in cement mortar (1:3) and flush-pointed as the work proceeds.

Note: (1) Brick copings to walls, sills, and steps are normally built entirely of facing bricks and therefore would be measured in accordance with S.M.M. G.22.1.

(2) It is optional to build up a price based on building 1,000 bricks, or $13\frac{1}{3}$ bricks which is, in this case, per linear metre. The writers believe that it is simpler to price on the basis of 1,000 bricks and reduce to the required unit at the end, as this taxes the memory less.

Cost of 1,000 facing bricks d/d site including stacking and waste	6,699p
Labour	
Based on 2 bricklayers and 1 labourer; based on 30 bricks per hour, ∴ Cost of building 1,000 = $\frac{1,000}{60} \times 630p$	10,500p
Mortar (Example 2): $0{\cdot}60\ m^3$ builds 1,000 bricks in cope 225 mm wide, ∴ $1,864p \times 0{\cdot}60$	1,118p
Cost of building 1,000 bricks in coping	18,317p
Number of bricks per lin.m of coping $13\frac{1}{3}$	
∴ $\frac{18,317p}{1,000} \times 13\frac{1}{3}$	244p
Bricklayer would point equivalent of $1{\cdot}50\ m^2$ per hour of cope	
1 lin.m of cope (which is 450 mm girth) = $0{\cdot}46\ m^2$	
∴ Cost of pointing 1 lin.m of cope = $\frac{220p}{1{\cdot}50} \times 0{\cdot}45$	66p
	310p
Profit and Oncost 20%	62p
	372p
Rate per m	£3·72

Note: The same labour factor would apply to brick-on-edge sills or steps.

Example 27

Brick-on-end flat arch of 65 mm facing bricks (Prime Cost d/d site £60·00 per 1,000) 225 mm high and 113 mm wide on soffit built on angle iron arch bar (measured separately) in cement mortar (1:3) and flush-pointed as the work proceeds. (Full value.)

Note:

(i) An arch such as this would possibly be built in the outer skin of a cavity wall built entirely of facings. In such a case it is measured as full value and not extra over.

(ii) S.M.M. G22 refers to facework to arches which are measured as extra over common brickwork when facing a wall of common brickwork.

(iii) It is suggested that the labour constants should be same as previous example but the mortar content would be slightly less due to being built in half-brick wall.

Cost of 1,000 facing bricks—as previous example		6,699p
Labour—as previous example		10,500p
Mortar: 0·53 m^3 builds 1,000 bricks, ∴ 1,864p x 0·53p		988p
Cost of building 1,000 bricks in flat arch		18,187p
Number of bricks per linear metre of flat arch =	$13\frac{1}{3}$	
∴ $\frac{18,187p}{1,000} \times 13\frac{1}{3}$ =		242p
Bricklayer pointing: total girth 338 mm, less angle iron 80 mm = 258 mm		
1 lin.m of arch is average 258 mm girth (say 0·26 m^2);		
∴ Cost of pointing 1 lin.m = $\frac{220p}{1{\cdot}50} \times 0{\cdot}26$		38p
	C/fd	280p

	B/fd 280p
Profit and Oncost 20%	56p
	336p
Rate per m	£3·36

BLOCKWORK

This term in the S.M.M. Sixth Edition refers to all types of blockwork, e.g. hollow clay, clinker or breeze concrete, moler, lightweight concrete, and glass.

Example 28

75 mm breeze block partition in blocks size 450 x 225 mm built in cement-lime mortar (1:1:6)

Breeze blocks are purchased by square metre and the S.M.M. unit of measurement is the square metre; therefore the rate is calculated on the basis of a square metre, the labour constant and mortar required being related to the unit.

Breeze blocks delivered site £1·50 per m^2	150p
Unload and stack; labourer 2 minutes @ 190p per hour	6p
	156p
Allow waste 5%	8p
Mortar including waste 0·007 m^3 @ 1,643p per m^3 (*see* Example 4)	12p
Labour based on 2 bricklayers and 1 labourer; output per hour per bricklayer = 1·50 m^2	
∴ Cost of building 1 m^2 = $\frac{630\text{p}}{3}$	210p
	386p
Profit and Oncost 20%	77p
	463p
Rate per m^2	£4·63

Example 29

100 mm Thermalite blockwork type
450 x 225 mm built in cement-lime m
partitions.

Thermalite blocks per m^2 d/d site
£2·05 per m^2
Unload and stack: labourer 2 minutes
@ 190p per hour

Allow waste 5%	p
Mortar, including waste, $0{\cdot}01\ m^3$ @ 1,643p per m^3	16p
Labour based on 2 bricklayers and 1 labourer; output per hour per bricklayer = $1{\cdot}25\ m^2$	
∴ Cost of building 1 m^2 is 1 hour @ $\frac{630p}{2{\cdot}50}$	252p
	490p
Profit and Oncost 20%	98p
	588p
Rate per m^2	£5·88

Example 30

Glass blockwork in panels 80 mm thick composed of 'panobrick' hollow glass blocks, built in cement-lime mortar (1:1:4). Type H 396 C, each size 190 x 190 x 80 mm thick.

Glass blocks are purchased by number, and are built to 200 mm courses.

Number of blocks per m^2 =

$$\frac{1{,}000 \times 1{,}000}{200 \times 200} = 25$$

25 blocks d/d site cost £1·55 each	3,875p
Unload and stack; 10 minutes labourer @ 190p	32p
C/fd	3,907p

		B/fd	3,907p
...ow waste 5%			195p
			4,102p
Mortar (*see* Example 3)			
1 m³ cement = 1·44 tonnes @ 2,552p	3,675p		
1 m³ lime = 0·60 tonne @ 2,952p	1,771p		
4 m³ sand = 6·08 tonnes @ 210p	1,277p		
	6,723p		
Shrinkage is 25%, ∴ add 33⅓%	2,241p		
6 m³ costs	8,964p		
∴ 1 m³ costs	1,494p		
Labourer mixing 5 hours @ 190p	950p		
Cost of 1 m³	2,444p		
Mortar including waste: 0·01 m³; ∴ 887p x 0·01			24p
Bricklayer builds about 0·80 m² per hour (20 blocks) including pointing			
Squad of 2 bricklayers and 1 labourer will produce 1·60 m² per hour @ 630p			
$\therefore 1\ m^2 = \frac{630p}{1 \cdot 60} =$			394p
			4,520p
Profit and Oncost 20%			904p
			5,424p
Rate per m²			£54·24

Example 31

Bitumen dampcourse on walls, overlapped 75 mm at all joinings.

Note: Sheet dampcourse is normally bought in rolls.

D.P.C. cost per m^2	250p
Allow for waste and laps 5%	13p
Bricklayer laying 15 minutes per m^2 @ 220p per hour	55p
Labourer laying 7½ minutes per m^2 @ 190p per hour	24p
	342p
Profit and Oncost 20%	68p
	410p
Rate per m^2	£4·10

Note: Sheet dampcourse is sold by the square metre for all widths, i.e. from widths 900 mm wide down to widths 112·5 mm wide, the cost of narrower widths being a straight proportion of the square metre price.

The S.M.M. Sixth Edition Clause G 38 states that dampcourse over 225 mm wide shall be given in square metres, which means that all widths exceeding 225 mm would be included in the above rate. For dampcourse 225 mm wide, the above square metre rate would merely be divided by 1,000/225 to give a rate of 92p per lin.m, and for dampcourse 100 mm wide the rate would be divided by 10 to give a rate of 41p per lin.m.

CENTERING

To flat soffit not exceeding 300 mm wide per m 0·33 hour carpenter

Timber required per m 0·01 m^3

For segmental work the labour constant should be doubled and the quantity of timber increased by 50%.

8 Underpinning

The labour output in underpinning is considerably lower than in normal working conditions as the work is carried out in confined spaces. Generally the output is about 50% of normal.

EXCAVATIONS

(hand excavation)

	Labourer hours per m^3
Trench in normal soil	5
Trench in heavy clay	6·66
One additional throw (1·50 m stage)	2

CONCRETE

The mixing of concrete would be the same as for concrete foundation; *see* Chapter 6, Examples 1 and 2 and Table of Transporting Times. The transporting and placing would be:

	Labourer hours per m^3
Foundations 150 to 300 mm thick	4
Foundations over 300 mm thick	3

For cutting away old concrete foundations and brick footings, by hand, allow 12 hours bricklayer per m^3. This volume is converted to linear dimensions as required; e.g. if total volume of projecting foundations and footings were 0·25 m^3 per lin.m, the time required would be 3 hours.

BRICKWORK

The bricks normally used for underpinning are load-bearing clay brick or blue engineering bricks (2nd quality).

The average output is:

Reduced brickwork 30 bricks per hour per tradesman
Wedging and pinning up to underside of existing concrete,
2·7 hours per tradesman per m^2

9 Rubble Walling, Masonry, and Cladding

RUBBLE WALLING

There is a large variety of types of rubble walling in various parts of the country, such as dry stone walling built random or in courses, walling of random rubble, coursed rubble and snecked rubble. Coursed rubble and snecked rubble may be built with roughly squared joints, or with dressed joints. The prices will also vary accordingly.

RUBBLE

The average weight of stone may be taken as approximately 60 kg per 25 mm of thickness of stone per square metre, i.e. 300 mm thickness of rubble walling would require 720 kg per m^2 and a 400 mm thickness of rubble walling would require 960 kg per m^2. These figures may be taken as inclusive of waste.

MORTAR

The quantity of mortar required per cube metre of walling varies according to the size of the stones and is approximately 0·20 m^3 for 75 mm stones and 0·13 m^3 for 300 mm stones.

An average of 0·17 m^3 for walling is reasonable.

LABOUR OUTPUT

The output is based on the cube metre and related to the required thickness.

2 tradesmen and 1 labourer: 6 hours per m^3 for 75 mm stones;
2 tradesmen and 1 labourer: $5\frac{1}{4}$ hours per m^3 for 300 mm stones.

Example 1

Rubble walling 450 mm thick built of second-hand sandstone

to snecked rubble pattern with roughly squared joints in cement-lime mortar 1:1:6 and pointed both sides as work proceeds.

1 m² of second-hand rubble 450 mm thick requires $\frac{60 \times 450}{25} = 1{,}080$ kg	
1,080 kg @ £7·50 per tonne d/d site (rubble is tipped from lorry)	810p
Mortar for building 0·17 m³ @ £21·77 per m³ (Chapter 7, Example 3)	370p
2 tradesmen and labourer build 1 m³ in 5·25 hours	
∴ time required for 1 m², 450 mm thick = $5{\cdot}25 \times \frac{450}{1{,}000} = 2{\cdot}36$ hours	
2·36 hours @ 630p	1,487p
Pointing requires ½ hour per face = for 2 faces 1 hour;	
∴ 1 tradesman and ½ labourer; 1 hour @ 315p	315p
	2,982p
Profit and Oncost 20%	596p
	3,578p
Rate per m²	£35·78

Example 2

Rubbed sandstone ashlar facework in courses 125 mm broad and 300 mm high in outer skin of hollow wall built in cement-lime mortar (1:1:6) and raked out and pointed on face with cement mortar 1:3.

Material

Red sandstone ashlar ready rubbed cut and jointed (in random lengths) £18·50 per m²	1,850p
Unload and stack ½ hour labourer; @ 190p	95p
C/fd	1,945p

B/fd	1,945p
Mortar for bedding 0·014 m³ @ £21·77 (*see* Chapter 7, Example 3)	30p
Mortar for pointing, nominal say	6p
	1,981p
Allow for waste 10% (this is for cutting random lengths to size)	198p
	2,179p
Labour	
Tradesman requires 2½ hours per m² to build ashlar;	
∴ 2 tradesmen and 1 labourer = 1¼ hours @ 630p	788p
1 tradesman and labourer cleaning down and pointing 1½ hours @ 410p	615p
	3,582p
Profit and Oncost 20%	716p
	4,298p
Rate per m²	£42·98

Example 3

300 x 150 mm Yorkshire sandstone sill rubbed on exposed face, sunk weathered, throated, and grooved.

Consider a sill 1·50 long = Cost per m £23·00 per m. (*Note:* The stated cost is for sill ready worked to specification.)

A 300 x 150 mm sill 1·50 m long @ £23·00 per m	3,450p
Unload and stack labourer ¼ hour @ 190p	48p
Mortar for bedding and pointing, nominal	6p
	3,504p
Allow 2½% waste for breakages	88p
C/fd	3,592p

B/fd	3,592p
Tradesman and labourer hoisting and building, 1 hour @ 410p	410p
	4,002p
Profit and Oncost 20%	800p
Cost of sill	4,802p
Rate per m = 4,802p x $\frac{2}{3}$	£32·01

Example 4

300 x 150 mm Yorkshire sandstone cope rubbed on exposed face, weathered and twice throated.

300 x 150 mm cope, cost per m £19·00	1,900p
Unload and stack; labourer 10 minutes @ 190p	32p
Mortar, nominal, say	4p
	1,936p
Allow $2\frac{1}{2}$% waste (for breakages)	48p
	1,984p
Tradesman and labourer hoisting and building, $\frac{1}{2}$ hour @ 410p	205p
	2,189p
Profit and Oncost 20%	438p
	2,627p
Rate per m	£26·27

CLADDING

Example 5

38 mm Westmorland slate cladding to walls in slabs size 450 x 900 mm with finely rubbed face, set risband on brick walls on cement mortar dots with fine joints, and with two phosphor bronze cramps and dowels to each horizontal bed

and vertical joints, including cutting brickwork for cramps, and drilling slate for dowels and building in cramps and pointing cladding.

Material

1 m^2 of slate cladding, cut to size, finely rubbed and drilled for dowels	4,560p
Unload and stack; labourer unloads 18 m^2 per hour @ 190p	11p
Number of cramps required per m^2 at 4 per slab size 450 x 900 = $\frac{1,000 \times 1,000}{450 \times 900} \times 4$ = 9·88 cramps—say 10 @ 30p each	300p
Mortar (1:3) dots at 4 per slab = 10 per m^2 and for bedding and pointing, allow	28p
	4,899p
Waste 5%	245p
	5,144p

Labour

1 tradesman with power drill will cut 6 pockets for cramps per hour. Allow 5p per hour for use of drill = total of 225p per man and drill ∴ Cutting 10 pockets at 225p per hour = 225p x $\frac{10}{6}$	375p
2 tradesmen to 1 labourer = 2 x 220p + 190p = 630p	
1 tradesman will set and build 0·75 m^2 per hour @ 315p = 315p ÷ 0·75	420p
	5,939p
Profit and Oncost 20%	1,188p
	7,127p
Rate per m^2	£71·27

10 Roofing

TILES

Plain tiling

The materials for plain tiling are either clay or, in increasing quantity, concrete.

267 x 164 mm holed and nibbed is the standard tile but 255 x 155 mm holed only and 280 x 180 mm holed and nibbed are available.

Single-lap tiling

Single-lap tiles are so termed because the single overlap of one tile upon another is relied upon for weathertightness, whereas with plain tiles double lapping is necessary.

TYPES IN USE (NOT EXHAUSTIVE)

Type	Size
interlocking bold roll pantile interlocking	420 x 332 mm
double Roman interlocking double pantile	430 x 382 mm
English pantile	420 x 332 mm
interlocking slate	430 x 380 mm
plain and cross-cambered	267 x 164 mm

SLATES

Roofing slates come from:

(1) Wales: Bangor and Caernarvon, Portmadoc, Corris (sized slates) sold by 1,000 and Precelly (random sized).

(2) Scotland: Ballachulish, Balvicar and Ullapool were sold unsized by 1,000. Full size 355 x 203 mm. Undersize 254 x 152 mm. Quarries now closed. Large quantities of second-hand slates available from demolitions.

(3) Cornwall: Old Delabole (random and sized slates) sold by tonne.

(4) Westmorland: Broughton Moor, Buttermere, and Tilberthwaite (random sized) sold by tonne.

(5) North Lancs.: Burlington (random and sized slates) sold by tonne.

Outputs for 1 slater and ½ labourer

Note: It is understood that the areas are big enough to warrant the time of a labourer being shared between two tradesmen.

Type	m^2 per hour
Slates	
305 x 152 mm to 330 x 178 mm	2¼
330 x 254 mm to 356 x 203 mm	2½
356 x 254 mm to 405 x 305 mm	3
457 x 229 mm to 508 x 305 mm 559 x 280 mm to 610 x 356 mm	4
Undersized random	1
Asbestos cement 610 x 305 mm	5
Labours (tradesman only)	
Angle cutting at valleys	7 m
Angle cutting at piends (hips)	14½ m
Mitred hips	1½ m
Ridging	5½ m
Straight cutting	6 m
Plain Tiles	
267 x 164 mm with 63 mm lap	2¼
Labours (tradesman only)	
Angle cutting at valleys	6 m
Angle cutting at piends (hips)	10½ m
Straight cutting	5 m
Interlocking, *etc. Tiles*	3¾
Labours (tradesman only)	
Angle cutting at valleys	4¼ m
Angle cutting at piends (hips)	10½ m

Underfelt	3 rolls
Battens	30 m

Waste: Allow 5% waste for unsized slates and $2\frac{1}{2}$% for sized slates and tiles.

Example 1

Single-ply bituminous underslating felt having 75 mm side laps and 150 mm end laps.

Basic Rates
Felt d/d site in 20 roll lots, each roll weighing 9·08 kg and approx. 10 m^2 in area @ £4·60 per roll
Nails 60p per kg

Material

1 roll of felt costs		460p
Unload and stack; 5 minutes @ 190p per hour		16p
		476p
Nails: Length of sheet	10·000 m	
Width of sheet	1·000 m	
	11·000 m with nails at 230 mm centres	
∴ No. of nails	$\frac{11{,}000}{230} = 48$	
770 nails weigh 1 kg @ 60p;		
∴ 48 nails cost $\frac{48}{770}$ x 60p		4p
		480p
Allow 5% waste		24p

Labour

1 tradesman and $\frac{1}{2}$ labourer will fix 3 rolls per hour		
∴ 1 tradesman @ 220p	220p	
C/fd	220p	504p

B/fd	220p	504p
1 labourer half-time @ 190p	95p	
3 rolls	315p	
∴ 1 roll		105p
Cost of 1 roll is		609p

1 roll covers 10·000   1·000
less   0·150   0·075

9·850 m × 0·925 m = 9·11 m²

∴ 1 m² is $\frac{609p}{9{\cdot}11}$	67p
Profit and Oncost 20%	13p
	80p
Rate per m²	80p

Example 2

Raking cutting on roofing felt.

Material	
Average waste is 0·1 m² per m of cut	
∴ waste = 0·1 of 67p (from Example 1, as felt has to be laid before being cut)	$6\frac{1}{2}$p
Labour	
1 slater will cut 30 m per hour @ 220p	
∴ 1 m is	$7\frac{1}{2}$p
	14 p
Profit and Oncost 20%	3 p
Rate per m	17 p

SLATE EXAMPLES

Example 3

Secondhand West Highland slates average 356 mm x 203 mm (14 x 8″) laid with 76 mm cover and single head nailed with 38 mm galvanized nails.

Material	
West Highland slates d/d site £215·00 per 1,000	21,500p
Take delivery and stack $1\frac{1}{2}$ hours per 1,000 @ 190p	235p
Nails: 1,000 per 1,000 slates @ 321 per kg;	
∴ 1,000 weigh 3·175 kg @ 55p per kg	175p
	21,910p
Labour	
Preparing slates	
Re-boring 4 hours	
Re-dressing 2 hours	
6 hours of tradesman @ 220p	1,320p
Sizing 2 hours of labourer @ 190p	380p
	23,610p
Allow 5% waste	1,180p
	24,790p

Covering capacity of 1 slate = Gauge x Width

$$\text{Gauge} = \frac{356\text{ mm} - (76 + 25\text{ mm})}{2} = 128\text{ mm}$$

∴ 128 x 203 mm = 0·0258 m²

∴ Covering capacity of 1,000 slates = 0·0258 m² x 1,000 = 25·8 m²

∴ Cost per m² is $\frac{24{,}790\text{p}}{25{\cdot}8}$ 961p

C/fd 961p

B/fd 961p

Labour
Covering roof, etc.
1 tradesman and $\frac{1}{2}$ labourer will fix $2\frac{1}{2}$ m^2 per hour;
$\therefore$ 1 m^2 costs $\frac{315\text{p}}{2\frac{1}{2}}$ 126p

1,087p
Profit and Oncost 20% 217p

1,304p

Rate per m^2 £13·04

Example 4

Extra for forming double eaves course of slates (using 356 x 203 mm West Highland slates).

Material
Slates cost 24,790p per 1,000 (*see* Example 3) per 100 m of eaves
493 slates (493 x 0·203 mm = 100 m) cost
$\frac{24{,}790}{1{,}000} \times 493 =$ 12,222p

Labour
From Example 3: 1,000 slates cover 25·8 m^2 and 1 tradesman and $\frac{1}{2}$ labourer will fix $2\frac{1}{2}$ m^2 per hour
No. of slates per m^2 = $\frac{1{,}000}{25 \cdot 8}$ = 39 say
39 slates take $\frac{60 \text{ minutes}}{2\frac{1}{2} \text{ m}^2}$ = 24 minutes
$\therefore$ 493 slates = $\frac{24}{39} \times 493$ = 303 minutes
say 5 hours @ 315p per hour 1,575p

13,797p
Profit and Oncost 20% 2,759p

100 m costs 16,556p

Rate per m £1·66

Example 5

Welsh slates 405 x 205 mm (16 x 8″) laid with 75 mm cover and single-head nailed.

Material			
Welsh slates d/d station £165·00 per 1,000			16,500p
say site 10 miles from station			
1,000 slates weigh approx. 1¾ tonnes			
Say lorry carries 7 tonnes, ∴ load is 4,000 slates			
Load lorry @ 1½ hours per 1,000	6 hours		
Unload and stack @	6 hours		
Travelling time 4 men @ ½ hour each way	4 hours		
	16 hours		
@ 190p =		3,040p	
Lorry time say 5 hours (travelling and standing) @ £4·00 per hour (inc. extra cost of driver's over labourer's rate)		2,000	
4,000 d/d site cost		5,040p	
∴ 1,000			1,260p
Nails as Example 3: 1,000			175p
Labour			
Preparing slates			
Boring 4 hours @ 220p			880p
			18,815p
Allow 2½% waste			470p
			19,285p

Covering capacity

Gauge = $\frac{405 - (75 + 25 \text{ mm})}{2} = 153$

$\therefore$ 1 slate = 153 x 205 mm = 0·0314 m^2

$\therefore$ 1,000 covers 0·0314 m^2 x 1,000 = 31·4 m^2

$\therefore$ Cost per m^2 is $\frac{19{,}285}{31 \cdot 4}$ 614p

Labour

Covering roof, etc.

1 tradesman and $\frac{1}{2}$ labourer will fix 3 m^2 per hour;

$\therefore$ 1 m^2 costs $\frac{315}{3}$ 105p

	719p
Profit and Oncost 20%	144p
	863p
Rate per m^2	£8·63

Example 6

Extra for forming double course of slates at eaves using 405 x 205 mm Welsh slates.

Material

Slates cost 19,285p per 1,000 (*see* Example 5)	19,285p
Welsh slates required to be dressed to size 2 hours per 1,000 @ 220p	440p
	19,725p

per 100 m of eaves

488 slates (488 x 0·205 m = 100 m) cost

$\frac{19{,}725}{1{,}000} \times 488$ 9,626p

C/fd 9,626p

	B/fd	9,626p

Labour
From Example 5, 1,000 slates cover 31·1 m^2 and 1 tradesman and ½ labourer will fix 3 m^2 per hour

No. of slates per m^2 = $\frac{1,000}{31 \cdot 1}$ = 32 say

32 slates take $\frac{60 \text{ minutes}}{3 \text{ m}^2}$ = 20 minutes

∴ 493 slates = $\frac{20}{32}$ x 493 = 308 minutes

say 5 hours 10 minutes @ 315p per hour	1,628p
	11,254p
Profit and Oncost 20%	2,251p
100 m costs	13,505p
Rate per m	£1·35

Note: Example 6 should not be taken as a follow-up to Example 5.

Example 7

Extra for forming verge on slates.

Material

Cost of 405 x 305 mm slates d/d station £250·00 per 1,000	25,000p
Haulage 1,260p (from Example 5) plus 50% (for size)	2,090p
Nails: double-nailed (*see* Example 5)	350p

Labour

Boring holes 6 hours per 1,000 for double holing @ 220p	1,320p
	28,760p
Allow 2½% for waste	719p
1,000 cost	29,479p

Per 100 m of verge
Gauge is 153 mm (Example 5) with
405 x 305 mm slate every 2nd course
∴ No. of slates per 100 m = 100,000 ÷ (153
x 2) = 327 say of 405 x 305 mm

∴ 327 slates @ 29,479p per 1,000	9,640p
less 490 slates (405 x 205 mm) @ 19,285p per 1,000	9,450p
	190p

Note: 405 x 305 mm slates cover $1\frac{1}{2}$ times the area of 405 x 205 mm slates.

Labour

Straight cutting 17 hours per 100 m @ 220p per hour	3,740p
	3,930p
Profit and Oncost 20%	786p
100 m costs	4,716p
Rate per m say	47p

Example 8

Square cutting slates at large openings.

Material
Waste is included in slate item

Labour

Straight cutting 6 m per hour @ 220p	37p
Profit and Oncost 20%	7p
Rate per m	44p

Example 9

Westmorland Best Peggies 305 to 229 mm long laid with 76 mm cover at eaves diminishing to 51 mm cover at ridge, single head nailed.

Material

Cost of slates d/d site £152·00 per tonne	15,200p
Unload and stack; 1 tonne per hour @ 190p	190p
Cost per tonne	15,390p
No. of slates per tonne = 1,690 (average)	
Nails @ 3·175 kg per 1,000 = 5·37 kg @ 55p per kg	295p
Labour	
Preparing slates	
Boring 7 hours	
Dressing 3½ hours	
10½ hours of tradesman @ 220p	2,310p
	17,995p
Sizing, 3½ hours labourer @ 190p	665p
	18,660p
Allow 5% for waste	933p
Cost per tonne	19,593
1 tonne covers 20 m^2	
∴ Cost of slates per m^2 = 19,593 ÷ 20	980p
Labour	
Covering roof etc.: 1 tradesman and ½ labourer will fix 1 m^2 per hour @ 315p	315p
	1,295p
Profit and Oncost 20%	259p
	1,554p
Rate per m^2	£15·54

TILE EXAMPLES

Example 10

267 x 164 mm concrete plain tiles laid with 63 mm lap including 50 x 25 mm redpine battens, each tile single-nailed.

Material		
Tiles d/d site £78·00 per 1,000	7,800p	
Unload and stack (1,000 weigh $1\frac{1}{3}$ tonnes) @ 1 tonne per hour @ 190p	253p	
Nails: 1,000 nails weigh 3·175 kg @ 55p per kg	175p	
	8,228p	
Allow $2\frac{1}{2}$% for waste	206p	
	8,434p	

Covering capacity of tile = $\frac{267 - 63}{2} \times 164$

= 0·0167 m²

∴ Covering capacity of 1,000 tiles = 16·7 m²		
∴ Cost per m² is 8,434p ÷ 16·7		505p
50 x 25 mm battens: amount in 1 m² 1,000 ÷ 102 mm (gauge) = 10 rows each 1 m long = 10 m		
10 m @ 20p =	200p	
Nails: 0·09 kg @ 55p per kg (0·9 kg per 100 m)	5p	
	205p	
Allow 5% for waste	10p	215p
		720p
Labour		
Battens		
1 tradesman and $\frac{1}{2}$ labourer will fix 30 m per hour @ 315p ∴ 10 m =		105p
	C/fd	825p

	B/fd	825p
Tiles		
1 tradesman and $\frac{1}{2}$ labourer will fix $2\frac{1}{4}$ m^2 per hour		
$\therefore$ 1 m^2 costs $\frac{315}{2\frac{1}{4}}$		140p
		965p
Profit and Oncost 20%		193p
		1,158p
Rate per m^2		£11·58

Note: Slaters employed in fixing sand-faced roofing tiles receive 1p extra per hour.

Example 11

380 x 230 mm interlocking concrete tiles laid with 75 mm lap on 50 x 25 mm red pine battens, each tile single-nailed.

Material	
Tiles d/d site £127·00 per 1,000 =	12,700p
Unload and stack (1,000 weigh 2·9 tonnes, @ 1 tonne per hour @ 190p	551p
Nails	
50 mm nails weighing 5·44 kg per 1,000 @ 60p per kg	326p
	13,577p
Allow $2\frac{1}{2}$% for waste	340p
	13,917p

Covering capacity is gauge × width		
= 305 × 204 mm (26 mm side lap)		
= 0·0622 m^2		
∴ 1,000 tiles cover 62·2 m^2		
∴ Cost per m^2 is 13,917p ÷ 62·2		224p
50 × 25 mm battens: amount in 1 m^2		
1,000 ÷ 305 mm (gauge) = $3\frac{1}{3}$ rows each 1 m long = $3\frac{1}{3}$ m		
$3\frac{1}{3}$ m @ 20p	67p	
Nails: 0·03 kg @ 60p per kg	2p	
	69p	
Allow 5% for waste	3p	72p

Labour

Battens	
1 tradesman and ½ labourer will fix 30 m per hour @ 315p, ∴ $3\frac{1}{3}$ m	35p
Tiles	
1 tradesman and ½ labourer will fix $3\frac{3}{4}$ m^2 per hour, ∴ 1 m^2 costs $\frac{315p}{3\frac{3}{4}}$	84p
	415p
Profit and Oncost 20%	83p
	498p
Rate per m^2	£4·98

Example 12

343 × 242 mm clay pantiles laid on and including battens to 76 mm cover and 38 mm side lap, each tile single-nailed.

Materials

Tiles d/d site £312·00 per 1,000	31,200p
Unload and stack (1,000 weigh 2 tonnes) @ 1 tonne per hour @ 190p	380p
C/fd	31,580p

B/fd	31,580p	
Nails: 38 mm nails weighing 3·175 kg per 1,000 @ 55p per kg	175p	
	31,755p	
Allow $2\frac{1}{2}$% for waste	794p	
	32,549p	

Covering capacity of tile = gauge x width = 267 x 204 mm = 0·0545 m^2		
∴ 1,000 tiles cover 54·5 m^2		
∴ Cost per m^2 is 32,549p ÷ 54·5		597p
50 x 25 mm battens: amount in 1 m^2		
1,000 ÷ 267 mm (gauge) = $3\frac{3}{4}$ rows each 1 m long		
∴ $3\frac{3}{4}$ m @ 20p	75p	
Nails: 0·04 kg @ 60p per kg	$2\frac{1}{2}$p	
	$77\frac{1}{2}$p	
Allow 5% for waste	$3\frac{1}{2}$p	81p
Labour		
Battens		
1 tradesman and $\frac{1}{2}$ labourer will fix 30 m per hour @ 315p, ∴ $3\frac{3}{4}$ m		39p
Tiles		
1 tradesman and $\frac{1}{2}$ labourer will fix $3\frac{3}{4}$ m^2 per hour, ∴ 1 m^2 costs $\frac{315p}{3\frac{3}{4}}$		84p
		801p
Profit and Oncost 20%		160p
		961p
Rate per m^2		£9·61

Example 13

Extra for forming double course of tiles at eaves.

Materials		
164 x 164 mm eaves tiles d/d site £80·00 per 1,000	8,000p	
Unload and stack (1,000 weigh $\frac{3}{4}$ tonne) @ 1 tonne per hour @ 190p	143p	
Nails: 38 mm nails weighing 3·175 kg per 1,000 @ 55p per kg	175p	
	8,318p	
Allow $2\frac{1}{2}$% for waste	208p	
1,000 cost	8,526p	
Per 100 m of eaves		
610 tiles (610 x 0·164 mm = 100 m) cost $\frac{8{,}526p}{1{,}000}$ x 610		5,201p
50 x 25 mm red pine battens for 100 m @ 20p per m	2,000p	
50 mm nails (0·75 kg per 100 m) @ 60p per kg	45p	
	2,045p	
Allow 5% for waste	102p	2,147p
		7,348p
Labour		
Battens		
1 tradesman and $\frac{1}{2}$ labourer will fix 30 m per hour @ 315p ∴ 100 m costs		1,050p
	C/fd	8,398p

B/fd	8,398p
Tiles	
1 tradesman and ½ labourer will fix 80 tiles per hour ∴ 610 tiles cost $\frac{315p}{80}$ x 610	2,402p
	10,800p
Profit and Oncost 20%	2,160p
100 m costs	12,960p
Rate per m	£1·30

Example 14

Angle cutting concrete plain tiles at valleys.

Material	
Average waste is 0·1 m^2 per m of cut ∴ waste = 0·1 of 715p (*see* Example 10) (720p less 5p for nails)	71½p
Labour	
1 tiler will cut 6 m per hour @ 220p, ∴ 1 m	36½p
	108p
Profit and Oncost 20%	22p
Rate per m	£1·30

Example 15

Extra for forming verges on concrete plain tiles, pointed in cement mortar.

Material	
Tile and half d/d site £180·00 per 1,000	18,000p
Unload and stack (1,000 weigh approx. 2¼ tonnes) @ 1 tonne per hour @ 190p	428p
Nails: Double-nailed (*see* Example 10) 2 x 175p	350p
C/fd	18,778p

	B/fd	18,778p
Allow $2\frac{1}{2}\%$ for waste		470p
		19,248p
In 100 m of verge with gauge of 102 mm No. of tile and half = 100 m ÷ 204 mm (every 2nd course) = 490		
∴ 490 tiles and half @ 19,248p per 1,000		9,432p
Less 735 plain tiles @ 8,434p per 1,000		6,199p
		3,233p
Mortar		
0·03 m³ per 100 m @ 1,448p per m³ (hand-mixed due to small quantities involved) (Chapter 16, Example 9)		43p
Labour		
1 tradesman and ½ labourer will do 6 m per hour of pointing @ 315p ∴ 100 m		5,250p
1 tradesman will do 6 m per hour of straight cutting @ 220p ∴ 100 m		3,667p
		12,193p
Profit and Oncost 20%		2,439p
100 m costs		14,632p
Rate per m		£1·46

Example 16

Extra for saddleback ridge tiles in 457 mm lengths with 152 mm wings bedded and pointed in cement mortar tinted to match.

Material		
10 ridge tiles d/d site £5·90		590p
Unload and stack @ $2\frac{1}{2}$p each		25p
		615p
Allow $2\frac{1}{2}\%$ for waste		15p
	C/fd	630p

	B/fd	630p
Mortar		
0·014 m³ per 10 tiles (*see* Brickwork, Example 16), 4,806p per m³ for tinted mortar		
∴ Cost is 0·014 m³ @ 4,806p per m³		67p
10 cost		697p
∴ 1 m costs 697p ÷ 4·8 (10 tiles cover approx. 457 x 10 plus joints, say 4·80 m)		145p
Labour		
1 tradesman and ½ labourer will fix 5½ m per hour @ 315p ∴ 1 m		57p
		202p
Profit and Oncost 20%		40p
Rate per m		£2·42

ASBESTOS CEMENT CORRUGATED SHEETING

This is sold by the square metre in standard lengths from 1·220 to 3·050 m (4′ 0″ to 10′ 0″) in 152 mm (6″) stages. The sheets usually have 152 mm (6″) end laps and side laps of 1½ corrugations.

Type of sheet	Width	Net covering width
Standard	762 mm (2′ 6″)	660 mm (2′ 2″)
Big six	1·086 m (3′ 6¾″)	1·035 m (3′ 4¾″)
Trafford tile	1·118 m (3′ 8″)	1·016 m (3′ 4″)

Waste 2½% to 5%.

Laps Although normally 1½ corrugations, can vary.

Waste Allowance

(1) If a 2·591 x 1·118 m Trafford Tile is specified net area covered would be

Area of sheet $2{\cdot}591 \times 1{\cdot}118$ m = $2{\cdot}897$ m^2
Area covered $2{\cdot}439 \times 1{\cdot}016$ m = $2{\cdot}48$ m^2

$$\text{Percentage reduction in area} = \frac{2{\cdot}897 - 2{\cdot}480}{2{\cdot}48} \times 100 = 17\%$$

17% plus say 3% waste = 20%

(2) $1{\cdot}829 \times 0{\cdot}762$ standard sheet would cover

Area of sheet $1{\cdot}829 \times 0{\cdot}762$ m = $1{\cdot}39$ m^2
Area covered $1{\cdot}677 \times 0{\cdot}660$ = $1{\cdot}11$ m^2

$$\text{Percentage reduction in area} = \frac{1{\cdot}39 - 1{\cdot}11}{1{\cdot}11} \times 100 = 25\%$$

25% plus say 5% waste = 30%

General

Corrugated asbestos sheets are fixed to wood purlins with drive screws and hook bolts are used when fixing to metal angles. On an average 5 drive screws or hook bolts are required per square metre, with an allowance of 5% for waste.

Output

For a squad of 2 tradesmen and 1 labourer:

	m^2 per hour
Normal slopes	7
Slopes over 50 degrees	6
Vertical	5

Labours

	1 tradesman per hour
Straight or raking cutting	5 lin.m
Circular cutting	2 lin.m
Eaves filling and barge board	5 lin.m
Unload and stack	90 m^2 per hour for 2 labourers

Example 17

Standard asbestos cement sheeting fixed to timber purlins on roof slopes not exceeding 50° at 838 mm centres with 76 x 6 mm galvanized drive screws, asbestos and lead washers (in 1·829 m lengths).

Material	
Sheeting d/d site £1·60 per m²	160p
Unload and stack 90 m² @ 380p (2 labourers)	4p
	164p
Waste and laps 30%	49p
	213p
5 drive screws and washers @ 4p	20p
5% waste on do.	1p
	234p
Labour	
2 tradesmen and 1 labourer @ 630p per hour	
7 m² per hour (carpenters receive 1p per hour extra tool allowance for asbestos sheeting)	90p
	324p
Profit and Oncost 20%	65p
	389p
Rate per m²	£3·89

BITUMINOUS FELT ROOFING

Supplied in various qualities in rolls of 10 m x 1 m (but can also be obtained on occasions in 15 m, 20 m and 30 m x 1 m rolls) and must conform to BS 747 with the bonding compound complying to BS 3940.

Waste 5%

Laps CP 144: Part 1: 1968 recommends lapped joints of at least 50 mm wide at the sides and 75 mm at ends of each length of felt.

Output

For 1 tradesman and 1 labourer per hour (including unloading):

1 layer of felt roofing nailed to boards	14 m^2
Fixing felt to bitumen on coverings up to 10°	10 m^2
Fixing felt to bitumen on coverings over 10° n.e. 50°	8 m^2
Fixing felt to bitumen on coverings over 50° and vertical	7 m^2
Fixing felt to bitumen on coverings in small areas	4 m^2
13 mm insulation board	25 m^2
Tarring roofs one coat (labourer only)	5 m^2
Spreading chips (labourer only)	20 m^2

Labours

Raking cutting 3 layers	8 m
Curved cutting 3 layers	4 m
Turning into groove and wedging	4 m
Welted drip	4 m

Laps

The adjustment for laps depends upon the specification.

(*a*) Take the minimum of 50 mm at the sides and 75 mm at the ends.

On a roll size of 1·00 x 10·00

Area of roll is 10·029 m^2

Net area = 0·950 x 9·925 = 9·43 m^2

$$\% \text{ for laps} = \frac{10{\cdot}00 - 9{\cdot}43}{9{\cdot}43} = 100 = 6\%$$

(*b*) If stated to be 75 mm at the sides and 150 mm at the ends.

Net area covered 0·925 x 9·850 = 9·11 m^2

$$\% \text{ for laps} = \frac{10{\cdot}00 - 9{\cdot}11}{9{\cdot}11} \times 100 = 10\% \text{ approx.}$$

An example:

(a) Waste and laps would be 5 + 6, say 11%.
(b) Waste and laps would be 5 + 10, say 15%.

Example 18

Three-layer built-up felt roofing on flat timber roofs composed of self-finished bitumen felt weighing 18 kg per 10 m^2 per roll laid with 75 mm side and 150 mm end laps finished with granite chips including insulation boarding under.

Material		
Felt @ £4·50 per 10 m^2;		
∴ 1 m^2	45p	
∴ 3 layers cost 45p x 3		135p
Allow 15% waste		20p
Insulation board per m^2	92p	
Primer £10·00 per 25 litres say	9p	
Bitumen $1\frac{1}{2}$ kg per layer @ 9p per kg = $1\frac{1}{2}$ x 9 x 3	40p	
Bedding compound	15p	
Granite chips £6·30 per tonne with coverage of 65 m^2 ∴ 1 m^2 costs	10p	
	166p	
Allow 5% waste	8p	174p

Labour		
1 tradesman and 1 labourer will produce 10 m^2 per hour of felt		
∴ 3 layers will cost $\frac{410\text{p} \times 3 \text{ layers}}{10 \text{ m}^2}$		123p
25 m^2 per hour of insulation = 410p ÷ 25		16p
1 labourer will produce		
5 m^2 per hour of tarring roof = 190p ÷ 5		38p
20 m^2 per hour of spreading chips = 190p ÷ 20		10p
	C/fd	516p

	B/fd	516p
Fuel, boiler, and haulage, say		20p
		536p
Profit and Oncost 20%		107p
		643p
Rate per m^2		£6·43

ASPHALT

Asphalt is normally supplied in 25·40 kg (56 lb) blocks.

Thickness of Asphalt	Amount reqd. in kg per m^2	1 tradesman and 1 labourer in m^2 per hour		
		Horizontal	Sloping	Vertical
One coat 13 mm thick	25	2·00	1·75	0·75
Two coats 19 mm thick	40	1·50	1·125	0·625
Two coats 25 mm thick	50	1·25	0·875	0·50

The above is for normal areas.
Allow 2½% waste.

Example 19

20 mm asphalt roofing to BS 988 laid to falls not exceeding 10° in two layers including sheathing felt underlay.

Materials

Felt per m^2	25p	
Asphalt 40 kg @ £37·00 per tonne	148p	
	173p	
Allow 2½% waste	4p	177p

Labour

1 tradesman and 1 labourer will produce 1·5 m^2 per hour

∴ 1 m^2 costs 410p ÷ 1·5 m^2	273p
C/fd	450p

	B/fd 450p
Fuel	
Cost of fuel, haulage, etc. taken as £7·00 per tonne of asphalt laid = 0·04 x £7·00	28p
	478p
Profit and Oncost 20%	96p
	574p
Rate per m^2	£5·74

Labours	1 tradesman and 1 labourer per hour
Forming channels up to 300 mm girth	1 m
Skirtings 150 mm high	1 m
Fillets	2 m

11 Metal Roofing

Sheet metal work to roofs, that is, lead, copper, aluminium, and zinc work, is normally done by plumbers. In some areas plumbers have mates as assistants, who are semi-skilled labourers. In most areas, however, the plumber is assisted by an apprentice whose stage of apprenticeship may vary from 1st year to 4th year. In view of this variety of assistance, output times will vary in accordance with the type of assistant the plumber may have. Therefore, for the purposes of estimating, the output of an apprentice will vary in relation to the amount he is paid. That is, an apprentice receiving $33\frac{1}{3}\%$ of a plumber's wage will be expected to have an output of $33\frac{1}{3}\%$ of the plumber's output and apprentices receiving 50% and 75% of a plumber's wage will have outputs of 50% and 75% of a plumber's output. This may also be applied commensurately to a plumber's (semi-skilled) mate.

For example, if a specific job requires 6 hours of a plumber's time, the same job, if carried out by a plumber and an apprentice receiving 50% of a plumber's wage, would require 4 hours of plumber and apprentice's time. And the same job, if carried out by a plumber and an apprentice receiving $33\frac{1}{3}\%$ of plumber's wage, would require $4\frac{1}{2}$ hours of plumber and apprentice's time.

Therefore, all examples shown will be based on plumber's time only.

The plumbing industry in the U.K. agreed to a new system of grading which came into operation in Scotland on the first Monday in September 1970 and was followed by the rest of the U.K. shortly after that.

There are now three grades of plumber, viz:

Approved Plumber	– which equates to a qualified tradesman
Advanced Plumber	– which equates to a working foreman
Technician	– which equates to a general supervisor

The terms of pay and conditions for labourers/plumber's mates and apprentices are also included in the system.

A copy of the most recent promulgation dated 3 August 1976 is shown below.

NOTICE OF PROMULGATION

To the Industry in Scotland and Northern Ireland – Increase in Rates of Wages

The Joint Industry Board has agreed that on and as from Monday 9 August 1976 the recently announced Government/ T.U.C. Pay Guidelines shall be applied in the Industry as follows:

1. The present rates of wages and weekly supplement shall continue to apply. These are as follows:

	Hourly Rate	*Weekly Supp.*		*Hourly Rate*	*Weekly Supp.*
Technician	115½p	£6	1st year Apprentice	33p	£2
Advanced	100½p	£6	2nd year Apprentice	48p	£3
Approved	92½p	£6	3rd year Apprentice	59p	£3·75
Labourer	81½p	£6	4th year Apprentice	74½p	£4·80

An apprentice in his 3rd or 4th year with a City and Guilds Advanced Craft Certificate shall receive 83½p per hour and a weekly supplement of £5·40.

An apprentice serving the balance of apprenticeship beyond 4th year (see Clause 8 of Apprenticeship Scheme) without an Advanced Craft City and Guilds Certificate shall receive 77p per hour and a weekly supplement of £4·95 and such an apprentice with an Advanced Craft City and Guilds Certificate shall receive 83½p per hour and a Weekly Supplement of £5·40.

The Weekly Supplement is not to be included in the rate for the calculation of overtime nor in the rate for any other payment such as under an incentive bonus scheme but is only to be added to gross earnings for the week. Absences will be deductable at the rate of 1/5 of the Supplement for each day of absence.

2. *There shall be added to the weekly earnings as calculated above (including the weekly supplement of each graded plumber, labourer and apprentice) a further individual weekly earnings supplement of 5% of such weekly earnings (including the weekly supplement) subject to the following maximum and minimum amounts:*

Weekly earnings supplement	Graded Plumbers Labourers and Apprentices age 18 and over	Apprentices under age 18			
		1st Year $33\frac{1}{3}$%	2nd Year 50%	3rd Year $62\frac{1}{2}$%	3rd Year with C & G Advanced Craft Certificate 90%
	£	£	£	£	£
Maximum	4·00	1·33	2·00	2·50	3·60
Minimum	2·50	0·83	1·25	1·56	2·25

The individual weekly earnings supplement is not to be included in the rate for the calculation of overtime nor in the rate for any other payment such as under an incentive bonus scheme but is only to be added to gross earnings for the week. Absences will be deductable from the maximum and minimum amounts applying to the individual at the rate of 1/5 for each day of absence.

3. Operatives should be paid the value of holiday credits on their current cards for public and annual holidays up to and including the Summer Holiday 1977 without the £6 and the 5% additions. Account was taken of the likely increase in August 1976 when current values of stamps were fixed.

The above promulgation took into account Phase II of the Government/T.U.C. Pay Guidelines and no further promulgation will be made until the 1 August 1977. We cannot however, forecast what the outcome and terms of Phase III will be at the present time but anticipate that a rate of 10p per hour more for plumbers than the rates shown for other building crafts is realistic.

Estimates will be based on an "All-in" rate of 230p per hour for an Approved plumber.

SHEET LEAD WORK

Lead is purchased by weight; therefore it is more convenient to estimate output by weight, and to convert to the required unit of measurement afterwards.

Labour fixing sheet lead is generally based on lead weighing 24·39 kg per m^2 (Code No. 5, BS 1178: 1969 Metric Units).

Rates include for cutting, hoisting, and fixing in position.

Schedule of Labour Constants for Sheet Lead Work

Flat roofs	6 plumber hours per 50 kg
Gutters	8 do.
Valleys	6 do.
Cornices	8 do.
Ridges and hips	7 do.
Flashings to roof-lights	8 do.
Stepped flashings	10 do.
All other flashings and aprons	8 do.
Soakers (fixed by slater)	2 do.

Note: Adjustments:
for 19·52 kg per m^2 (Code No. 4) lead: add $\frac{3}{4}$ hour per 50 kg; for 29·29 kg per m^2 (Code No. 6) lead; deduct $\frac{1}{2}$ hour per 50 kg.

Labour
Dressing lead over glass = 6 m per hour
Batting to groove = 9 m per hour
Dressing over corrugated asbestos = 5 m per hour
Basic price of lead £10·20 per 50 kg
Basic price of scrap lead £3·40 per 50 kg

Waste: The wastage on lead is small and should be taken as 5% of the difference between the price of new and that of scrap lead.

Example 1

Code No. 5 (24·39 kg per m^2) sheet lead to flat roof.

Cost of 50 kg sheet lead		2,482p
	C/fd	2,482p

B/fd	2,482p	
Take delivery – 1·00 tonne per hour of tradesman = $\frac{50}{1,000} \times 230p$	12p	2,494p
Add for waste		
Difference between cost	2,482p	
and scrap value	1,250p	
5% on	1,232p	62p
		2,556p
Labour		
Cutting, hoisting, and fixing 6 hours plumber @ 230p per hour		1,380p
Cost per 50 kg		3,936p
∴ Cost per m² = $\frac{3,936p}{50} \times 24{\cdot}39$		1,920p
Profit and Oncost 20%		384p
		2,304p
Rate per m²		£23·04

Example 2

Code No. 5 (24·39 kg per m²) sheet lead lining to sloping side gutter of chimney stack; 450 mm girth dressing over tilting fillet and flashing plate.

50 kg lead a.b.	2,556p
Labour	
Cutting, hoisting, and fixing 8 hours plumber @ 230p per hour	1,840p
Cost per 50 kg	4,396p

$\therefore$ Cost per $m^2 = \frac{4{,}396}{50} \times 24{\cdot}39$	2,145p
$\therefore$ Cost per m 450 mm girth = 2,145p x 0·45	965p
Profit and Oncost 20%	193p
	1,158p
Rate per m	£11·58

Example 3

Code No. 6 (29·29 kg per m^2) sheet lead in parapet gutter.

50 kg lead a.b.	2,556p
Labour	
Cutting, hoisting, and fixing 7½ hours plumber @ 230p per hour	1,725p
(*Note:* ½ hour per 50 kg is deducted for 29·29 kg per m^2 lead.)	
Cost per 50 kg	4,281p
$\therefore$ Cost per $m^2 = \frac{4{,}281p}{50} \times 29{\cdot}29$	2,508p
Profit and Oncost 20%	502p
	3,010p
Rate per m^2	£30·10

Example 4

Code No. 4 (19·52 kg per m^2) sheet lead in flashings 150 mm girth with 100 mm laps including turning into groove and fixing with lead batts at not more than 125 mm centres.

50 kg lead a.b.	2,556p
Labour	
Cutting, hoisting, and fixing 8¾ hours plumber @ 230p per hour	2,013p
Cost per 50 kg	4,569p

∴ Cost per m² $= \frac{4{,}569\text{p}}{50} \times 19{\cdot}52$	1,784p
∴ Cost per lin.m 150 mm girth = 1,784p x 0·15	268p
Allow for laps of 100 mm per 2·00 m	13p
Allow 8 batts per m, say	10p
Plumber batting 9 m per hour = 230p ÷ 9	26p
	317p
Profit and Oncost 20%	63p
	380p
Rate per m	£3·80

SHEET OR STRIP COPPER WORK

Sheet or strip copper is purchased by weight but, as it weighs much less per unit of area than lead (No. 24 gauge weighs 4·90 kg per m²), the cost of the sheet copper is converted to cost per square metre prior to commencing estimating and the output is based on hours per square metre.

Copper is now designated by thickness: see Amendment Slip No. 1 published 8 February, 1972, to CP. 143: Part 12: 1970.

The following note shows S.W. Gauges and their equivalent thicknesses.

	Thickness mm
26 S.W.G.	0·45
24 S.W.G.	0·56
23 S.W.G.	0·60
22 S.W.G.	0·70

At the present time it is, therefore, thought appropriate to define S.W.G. and thickness when describing sheet or strip copper.

Schedule of Labour Constants for Sheet or Strip Copper Work of Nos. 23 and 24 Gauge

	Plumber hours per square metre
Flat roofs	2·50
Gutters	5·00
Valleys	3·75
Cornices	5·00
Ridges and hips	4·40
Stepped flashings	6·25
All other flashings and aprons	5·00
Soakers (fixed by slater)	1·00

Labour	
Dressing over glass	6 m per hour
Batting to groove	9 m per hour
Dressing over corrugated sheeting	6 m per hour
Clinking copper	6 m per hour
Close copper nailing (25 to 38 mm spacing)	2·5 to 3·5 m per hour
Open copper nailing (50 to 100 mm spacing)	5 to 6 m per hour

Basic rate of copper = 164p per kg
Weight of No. 23 gauge = 5·80 kg per m^2
Weight of No. 24 gauge = 4·90 kg per m^2
Cost of No. 23 gauge copper per m^2 = 164p x 5·80 = 951·2p
Cost of No. 24 gauge copper per m^2 = 164p x 4·90 = 803·6p

As copper prices have been subject to extreme fluctuation, the examples will be calculated on 951p per m^2 for No. 23 gauge, and 804p per m^2 for No. 24 gauge.

Example 5

No. 24 gauge (0·56 mm thick) strip copper on timber flat roof.

No. 24 g strip copper per m²	804p
Take delivery, say	3p
	807p
Waste 2½%	20p
	827p
Labour	
Cutting, hoisting, and fixing 2·5 hours plumber @ 230p	575p
	1,402p
Profit and Oncost 20%	280p
	1,682p
Rate per m²	£16·82

Example 6

No. 23 gauge (0·60 mm thick) strip copper lining to gutter behind chimney stack 550 mm girth including dressing over tilting fillet and flashing plate.

No. 23 g strip copper per m²	951p
Take delivery, say	3p
	954p
Waste 2½%	24p
	978p
Labour	
Cutting, hoisting, and fixing 5 hours plumber @ 230p	1,150p
Cost per m²	2,128p

∴ Cost per m 550 mm girth = $\frac{550}{1,000}$ × 2,128p	1,170p
Profit and Oncost 20%	234p
	1,404p
Rate per m	£14·04

Example 7

No. 23 gauge (0·60 mm thick) strip copper in flashings 300 mm broad with 100 mm laps including turning into groove and fixing with copper batts at 125 mm centres.

No. 23 g strip copper per m²	951p
Take delivery, say	3p
	954p
Waste 2½%	24p
	978p
Labour	
Cutting, hoisting, and fixing 5 hours plumber @ 230p	1,150p
Cost per m²	2,128p
∴ Cost per m 300 mm broad = $\frac{300}{1,000}$ × 2,128p	638p
Allow for laps of 100 mm per 2·00 m	32p
Allow 8 copper batts per m @ 1½p	12p
Plumber batting 9 m per hour = 230p ÷ 9	26p
	708p
Profit and Oncost 20%	142p
	850p
Rate per m	£8·50

Example 8

Close copper nailing at 25 mm centre.

Number of nails per 10 m = 400	
Number of nails per kg = 880	
$\therefore \frac{400}{880}$ nails @ 150p per kg	68p
Waste $2\frac{1}{2}\%$	2p
Labour	
Plumber does 2·5 m per hour,	
$\therefore$ 10 m requires 4 hours @ 230p	920p
10 m cost	990p
$\therefore$ Cost per m	99p
Profit and Oncost 20%	20p
Rate per m	£1·19

SHEET ALUMINIUM WORK

Aluminium sheeting may be used for roof covering in a similar manner to copper. The construction, and technique in fixing, is exactly the same as for copper. Super purity aluminium is normally used and the gauge of the metal is normally slightly heavier than for copper, that is, where Nos. 24 and 23 gauges may be chosen for copper roofing, Nos. 22 and 21 gauges would be used for aluminium.

The Schedule of Labour Constants for aluminium is exactly the same as for copper.

Basic rate of aluminium = £2·40 per kg
Weight of No. 21 gauge = 2·20 kg per m^2
Weight of No. 22 gauge = 1·93 kg per m^2
Cost of No. 21 gauge aluminium per m^2 = 240p x 2·20 = 528p
Cost of No. 22 gauge aluminium per m^2 = 240p x 1·93 = 463·2p

The examples will be calculated on 530p per m^2 for No. 21 gauge; 465p per m^2 for No. 22 gauge.

The thickness of the gauge will also be stated. See CP 143:

Example 9

No. 22 gauge (0·70 mm thick) aluminium on timber flat roof.

No. 22 g aluminium per m²	465p
Take delivery, say	3p
	468p
Waste 2½%	12p
	480p
Labour	
Cutting, hoisting, and fixing 2·5 hours plumber @ 230p	575p
	1,055p
Profit and Oncost 20%	211p
	1,266p
Rate per m²	£12·66

Example 10

No. 21 gauge (0·90 mm thick) aluminium lining to gutter behind chimney stack; 550 mm girth including dressing over tilting fillet and flashing plate.

No. 21 g aluminium per m²	530p
Take delivery, say	3p
	533p
Waste 2½%	13p
	546p
Labour	
Cutting, hoisting, and fixing 5 hours plumber @ 230p	1,150p
Cost per m²	1,696p

∴ Cost per m 550 mm girth = $\frac{550}{1,000} \times 1,696$p	933p
Profit and Oncost 20%	187p
	1,120p
Rate per m	£11·20

Example 11

No. 21 gauge (0·90 mm thick) aluminium lining to sloping valley gutter 560 mm girth dressed over tilting fillets with 150 mm intermediate laps.

No. 21 g aluminium per m^2	530p
Take delivery, say	3p
	533p
Waste $2\frac{1}{2}$%	13p
	546p
Labour	
Cutting, hoisting, and fixing 3·75 hours plumber @ 230p	863p
Cost per m^2	1,409p
Cost per m 560 mm girth = $\frac{560}{1,000} \times 1,409$p	789p
Allow for laps of 150 mm per 2·00 m	59p
	848p
Profit and Oncost 20%	170p
	1,018p
Rate per m	£10·18

SHEET ZINC WORK

Zinc may be used for roof covering in place of lead, copper, or aluminium. The technique in fixing is similar to that for copper and aluminium, but labour output is slightly lower.

Schedule of Labour Constants for No. 14 Gauge Zinc
(weighing 5·67 kg per m^2)

	Plumber hours per m^2
Flat roof	2·75
Gutters	5·50
Valleys	4·125
Cornices	5·50
Ridges and hips	4·90
Stepped flashings	6·90
All other flashings and aprons	5·50
Soakers (fixed by slaters)	1·00

Labour

Dressing over glass	5·5 m per hour
Batting to groove	9·0 m per hour
Dressing over corrugated sheeting	5·5 m per hour
Clinking	5·5 m per hour
Close nailing	2·5 to 3·5 m per hour
Open nailing	5·0 to 6·0 m per hour

Zinc is purchased in sheets 8′0″ x 3′0″ (2·438 x 0·914 m)
Cost per sheet (size 2·228 m^2) = £10·16
∴ Cost per m^2 = £4·56

Example 12

No. 14 gauge zinc to flat roof.

No. 14 g zinc per m^2	456p
Take delivery, say	3p
	459p
Waste 2½%	12p
	471p
C/fd	471p

B/fd	471p
Labour	
Cutting, hoisting, and fixing 2·75 hours plumber @ 230p	633p
	1,104p
Profit and Oncost 20%	221p
	1,325p
Rate per m²	£13·25

Example 13

No. 14 gauge zinc in ridge capping 600 mm girth with 150 mm laps and with 650 x 50 mm cleats at 450 mm centres, dressed over slates.

No. 14 g zinc per m²	456p
Take delivery, say	3p
	459p
Waste $2\frac{1}{2}$%	12p
	471p
Labour	
Cutting, hoisting, and fixing 4·90 hours plumber @ 230p	1,127p
Cost per m²	1,598p
Cost per m 600 mm girth = $\frac{600}{1,000} \times 1,598$	959p
Allow for laps of 150 mm per 2·00 m	72p
Allow for cleats say 2·2 cleats per m = $\frac{650 \times 50 \times 2{\cdot}2}{1,000,000} \times 471\text{p}$	34p
	1,065p
Profit and Oncost 20%	213p
	1,278p
Rate per m	£12·78

12 Woodwork (Carpentry)

OUTPUTS

Timber classification	Tradesman cutting, hoisting, and fixing m³ per hour
Wallplate	0·14
Sleeper joists	0·12
Upper floor joists (and flat roof joists)	0·08
Ceiling joists	0·08
Purlins and struts	0·05
Rafters and collars	0·05
Roof truss members	0·05
Allow 50% on labour for lengths not exceeding 1·25 m (where defined)	
Roof boarding	4 m^2

Average amount of nails required per m^3 = 1·6 kg (excepting wallplate).

Cost of Nails

These costs are based on 5 tonne ordering rate.

Description	Length	Gauge	No. per kg	Per 50 kg
Bright Round Wire Nails				
	mm	mm		
(1″ x 15)	25	1·60	1,780	£22·80
(1½″ x 13)	40	2·36	740	£20·20
(2″ x 11)	50	3·00	348	£18·75
(2½″ x 10)	65	3·35	229	£18·20
(3″ x 9)	75	3·75	152	£18·00
(3½″ x 8)	90	4·00	108	£17·80
(4″ x 6)	100	5·00	64	£17·40
(5″ x 5)	125	5·60	44	£17·10
(6″ x 4)	150	6·00	31	£17·00

Description	Length mm	Gauge mm	No. per kg	Per 50 kg
Bright Oval Brads				
(1")	25		2,222	£23·60
(1¼")	30		1,122	£22·15
(1½")	40		972	£21·00
(2")	50		436	£19·40
(2½")	65		229	£18·60
(3")	75		125	£18·20
(4")	100		64	£17·17
Bright Lost Head Nails				
4" x 7	100		79	£17·70
Galvd. Round Wire Nails				
(1" x 15)	25		1,620	£33·15
(1½" x 13)	40		673	£27·15
(2" x 11)	50		317	£25·40
(2½" x 10)	65		209	£24·65
(3" x 9)	75		138	£24·33
(3½" x 8)	90		98	£24·15
(4" x 6)	100		59	£23·70
(5" x 5)	125		40	£23·40
(6" x 4)	150		28	£23·10
Galvd. Oval Brads				
(1")	25		2,022	£33·95
(1¼")	30		1,021	£29·95
(1½")	40		884	£27·95
(2")	50		395	£26·05
(2½")	65		208	£25·05
(3")	75		114	£24·55
(4")	100		58	£24·00
Sherd Jagged Plasterboard Nails				
(1¼" x 12)	30	2·65	625	£21·35
(1½" x 12)	40	2·65	539	£20·90
Galvd. Clout Nails (Ord. Head)				
(1¼" x 12)	30	2·65	594	£22·25
(1½" x 12)	40	2·65	515	£22·25
(1¼" x 11)	30	3·00	486	£22·06
(2" x 12)	50	2·65	390	£21·24

Description	Length mm	Gauge mm	No. per kg	Per 50 kg
Galvd. Clout Nails (*Extra Large Head*) (*hot dipped*)				
(⅝″ x 11)	13	3·00	583	£32·00
(¾″ x 11)	20	3·00	528	£31·00
(1″ x 11)	25	3·00	400	£31·00
Galvd. Slate Nails				
(1¼″ x 10)	30		480	£22·00
(1½″ x 10)	40		437	£22·00
(2″ x 10)	50		347	£21·00
(2½″ x 10)	65		257	£20·50
Galvd. Gypklith (*Slab*) *Nails*				
(4″ x 10)	100		99	£32·00
Galvd. Lath Nails				
(1″ x 14)	25		1,562	£22·66
(1¼″ x 14)	30		1,056	£22·18
(1½″ x 14)	40		871	£21·60

Structural timbers are now measured by the linear metre in accordance with S.M.M. Sixth Edition. The Timber Trades Federation recommend that linear timber be sold in units of 100 lin.m, stating the scantling sizes.

Linear timber may also be sold by the cube metre (m³), stating the scantling sizes, where large quantities are involved. The object of stating labour output for structural timbers in cube metres is to eliminate unnecessary labour constants for differing scantling sizes of timber within the same classification. Alternatively, the output may be expressed in linear metres per hour for each of the scantling sizes as shown below:

ALTERNATIVE OUTPUTS

Classification	Scantling size	Tradesman m per hour
Wallplate	25 x 100	56·00
	50 x 100	28·00
	50 x 150	18·67
Sleeper joists	50 x 100	24·00
	50 x 125	19·20

Classification	Scantling size	Tradesman per hour
	50 x 150	16·00
Upper floor joists	50 x 175	9·15
	50 x 200	8·00
	50 x 225	7·20
	50 x 250	6·40
	63 x 225	5·73
	63 x 250	5·08
Ceiling joists	50 x 100	16·00
	50 x 125	12·80
	50 x 150	10·67
Purlins and struts	50 x 100	10·00
	50 x 150	6·66
	75 x 150	4·45
Rafters and collars	50 x 100	10·00
	50 x 125	8·00
	50 x 150	6·66
Roof truss members	100 x 300	1·67
	100 x 200	2·50
	100 x 100	5·00

Typical average prices of carcassing timbers as at May 1977:

Scantling size mm	White pine per m³	Red pine per m³
75 x 275	£114·00	–
75 x 250	£113·00	–
75 x 225	£112·00	£122·50
50 x 200	£108·00	£119·50
50 x 175	£107·50	£117·00
50 x 165 50 x 150	£107·50	£117·00
50 x 138 50 x 125	£107·30	£117·00
50 x 115 50 x 100	£107·25	£117·00

TIMBER TREATMENT

Timber impregnated under pressure with 4·00 kg 'Tanalith' salts per m³; extra cost is £8·90 per m³.

Note:

(1) That the rates are calculated to the nearest penny in all of the following examples, but some estimators prefer to calculate to one decimal point of a new penny and round off the final rate.

(2) Carpenters may do their own stacking and hoisting or labourers may be employed. This would depend on size and structure of company.

Example 1

50 x 100 mm red pine tanalized wallplate.

Material	
50 x 100 mm tanalized red pine per m³ costs £125·90	12,590p
Unload and stack (average weight 430 kg per m³): ½ hour labourer @ 190 p	95p
Nails 0·5 kg @ £17·80 per 50 kg	18p
	12,703p
Waste 5%	635p
	13,338p

In 1 m³ there are 200 m of 50 x 100 mm

$$\left(\frac{1{,}000{,}000}{50 \times 100}\right)$$

$\therefore$ Cost per m = $\frac{13{,}338\text{p}}{200}$	67p
Labour	
Tradesman fixed 0·14 m³ per hour	
$\therefore$ Cost per m = $\frac{220\text{p}}{200 \times 0{\cdot}14}$	8p
	75p
Profit and Oncost 20%	15p
	90p
Rate per m	90p

Example 2

50 x 125 mm white pine sleeper joist.

Material

50 x 125 mm white pine per m³ costs £107·30	10,730p
Unload and stack: ½ hour labourer @ 190p	95p
Nails (75 mm): 1·6 kg @ £18·00 per 50 kg	58p
	10,883p
Waste 5%	544p
	11,427p

In 1 m³ there are 160 m of 50 x 125 mm

$\therefore\ 1\text{ m} = \dfrac{11{,}427\text{p}}{160}$	71½p

Labour

Tradesman fixed 0·12 m³ per hour

$\therefore\ \text{Cost per m} = \dfrac{220\text{p}}{160 \times 0{\cdot}12}$	11½p
	83p
Profit and Oncost 20%	17p
	100p
Rate per m	£1·00

Alternative method of estimating

Material

50 x 125 mm white pine per 100 m £67·06	6,706p
Unload and stack: ⅓ hour labourer @ 190p	63p
Nails (75 mm): 1·0 kg @ £18·00 per 50 kg	36p
	6,805p
Waste 5%	340p
C/fd	7,145p

B/fd	7,145p
Labour	
Tradesman fixes 19·20 m per hour	
∴ Cost per 100 m = $220\text{p} \times \frac{100}{19{\cdot}20}$	1,146p
	8,291p
Profit and Oncost 20%	1,658p
Cost per 100 m	9,949p
Rate per m	£1·00

Example 3

50 x 200 mm white pine floor joists.

Material	
50 x 200 mm white pine per m³ costs £108·00	10,800p
Unload and stack: ½ hour labourer @ 190p	95p
Nails 1·6 kg @ £18·00 per kg	58p
	10,953p
Waste 5%	548p
	11,501p
In 1 m³ there are 100 m of 50 x 200 mm	
∴ $1\text{ m} = \frac{11{,}501\text{p}}{100}$	115p
Labour	
Tradesman fixes 0·08 m³ per hour	
∴ Cost per m = $\frac{220\text{p}}{100 \times 0{\cdot}08}$	28p
	143p
Profit and Oncost 20%	29p
	172p
Rate per m	£1·72

Example 4

50 x 150 mm white pine ceiling joists.

Material

50 x 150 mm white pine per m³ costs £107·50	10,750p
Unload and stack: ½ hour labourer @ 190p	95p
Nails 1·6 kg @ £18·00 per 50 kg	58p
	10,903p
Waste 5%	545p
	11,448p

In 1 m³ there are 133·33 m of 50 x 150 mm

$\therefore$ Cost per m $= \dfrac{11{,}448\text{p}}{133{\cdot}33}$ — 86p

Labour

Tradesman fixes 0·08 m³ per hour

$\therefore$ Cost per m $= \dfrac{220\text{p}}{133{\cdot}33 \times 0{\cdot}08}$ — 21p

	107p
Profit and Oncost 20%	21p
	128p
Rate per m	£1·28

Example 5

50 x 125 mm white pine rafters.

Material

50 x 125 mm white pine per m³ costs £107·30	10,730p
Unload and stack: ½ hour labourer @ 190p	95p
Nails (100 mm): 1·6 kg @ £17·40 per 50 kg	56p
	10,881p
Waste 5%	544p
	11,425p

In 1 m^3 there are 160 m of 50 x 125 mm

$\therefore$ 1 m = $\frac{11{,}425p}{160}$ 71p

Labour
Tradesman fixes 0·05 m^3 per hour

$\therefore$ Cost per m = $\frac{220p}{160 \times 0{\cdot}05}$ 28p

	99p
Profit and Oncost 20%	20p
	119p
Rate per m	£1·19

Example 6

Labour trimming 50 x 150 mm white pine ceiling joists to opening size 850 x 850 mm (*see* diagram opposite).

Carpenter will make 8 square-housed joints per hour;

$\therefore$ 6 joints require $\frac{3}{4}$ hour tradesman @ 220p	165p
Profit and Oncost 20%	33p
Rate each	£1·98

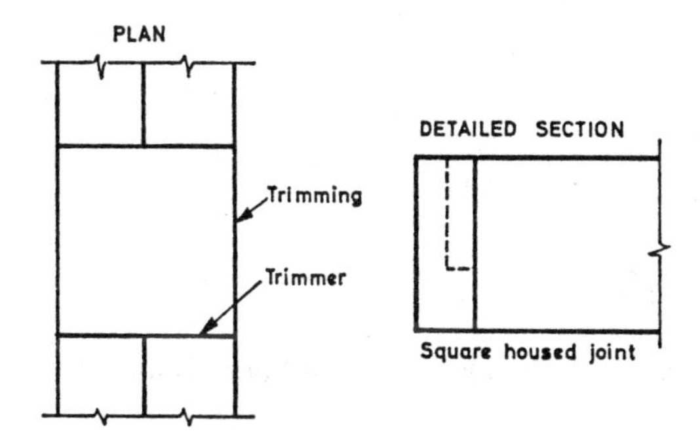

Example 7

38 x 200 mm white pine ridge.

Material	
38 x 200 mm white pine per 100 m £84·00	8,400p
Unload and stack: $\frac{2}{5}$ hour labourer @ 190p	76p
Nails 1·0 kg @ £18·00 per 50 kg	36p
	8,512p
Waste 5%	426p
	8,938p
Labour	
Tradesman fixes 8·00 m per hour	
∴ Cost per 100 m = 220p x $\frac{100}{8}$	2,750p
	11,688p
Profit and Oncost 20%	2,338p
Cost per 100 m	14,026p
Rate per m	£1·40

Note: Rate for 38 x 200 mm to hip rafter would be the same as above.

Example 8

50 x 200 mm white pine valley rafter.

Material	
50 x 200 mm white pine per 100 m £108·00	10,800p
Unload and stack: $\frac{1}{2}$ hour labourer @ 190p	95p
Nails 1·0 kg @ £18·00 per 50 kg	36p
	10,931p
Waste 5%	547p
	11,478p

Labour
Tradesman fixes 6·00 m per hour

∴ Cost per 100 m = 220p x $\frac{100}{6}$	3,667p
	15,145p
Profit and Oncost 20%	3,029p
Cost per 100 m	18,174p
Rate per m	£1·82

Example 9

15 mm white pine butt-jointed roof boarding to roof not exceeding 50° from horizontal; (assume rafters at 450 mm centres).

Material

Cost per m² £1·95	195p
Unload and stack	3p
Nails: Two 50 mm nails per 200 mm board per rafter;	

∴ In one m² there are 5·00 m of board; and with rafters at 450 mm centres;

Number of nails = $\frac{5{\cdot}00}{0{\cdot}45}$ x 2 = 22

22 nails @ £18·75 per 50 kg (160 per kg), say	6p
	204p
Waste 5%	10p

Labour
Tradesman fixes 4 m² per hour

∴ Cost per m² = $\frac{220p}{4}$	55p
C/fd	269p

	B/fd	269p
Profit and Oncost 20%		54p
		323p
Rate per m²		£3·23

Example 10

Raking, cutting on do.

Waste on cutting is 0·10 m² per m of cut

∴ Cost of waste = 198p x 0·10	20p
Labour	
Tradesman cuts 6 m per hour = $\frac{220p}{6}$	37p
	57p
Profit and Oncost 20%	11p
	68p
Rate per m	68p

Example 11

9·5 mm Gyproc Insulated Sarking (roof boarding) laid butt jointed to roof not exceeding 50° from horizontal with galvanized clout nails (assume rafters at 450 mm centres).

Note: Gyproc (Gypsum) Sarking is sold in units of 100 m² in sheet sizes 1,800 x 600 mm. It is marketed only in Scotland.

Materials		
Cost per 100 m² £60·80		6,080p
Unload and stack: 1½ hours labourer @ 190p		285p
Nails: 100 m² represents 91·1 sheets and with sheets set on rafters at 450 mm centres and at 150 mm centres; the number of nails required per sheet is 25		
∴ 2,278 are required per 100 m²		
There are 594 30 mm nails per kg;		
	C/fd	6,365p

B/fd	6,365p
∴ Nails required = $\frac{2,278}{594}$ = 3·8 kg @ £22·25 per 50 kg	169p
	6,534p
Waste 5%	327p
	6,861p

Labour

Tradesman fixes 4 m² per hour

∴ Cost per 100 m² = 220p x $\frac{100}{4}$	5,500
	12,361p
Profit and Oncost 20%	2,472p
Cost per 100 m²	14,833p
Rate per m²	£1·48

Where items such as open-spaced grounds and framework are measured by m² stating spacing of the members the length of the member per m² may be determined empirically as follows:

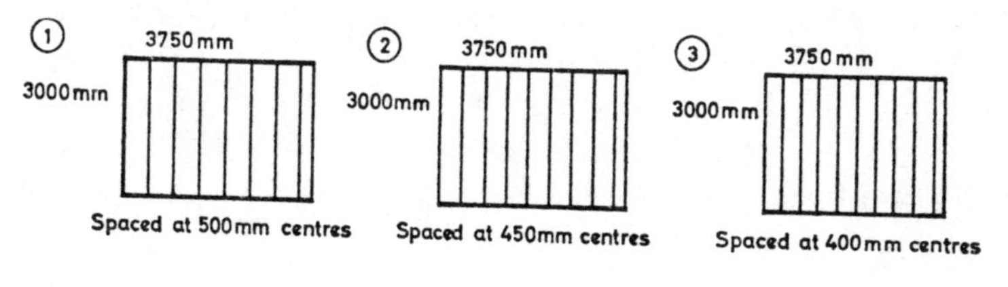

Area = 11·25 m² 11·25 m² 11·25 m²

Length of timber required:

9/3·00 = 27·00 m 10/3·00 = 30·00 m 11/3·00 = 33·00 m

Actual length of timber required per m² =

27·00 ÷ 11·25 = 2·40 m 30·00 ÷ 11·25 = 2·67 m 33·00 ÷ 11·25 = 2·94 m

To find length of timber on basis of member spacing:

$\frac{1,000}{500} \times 1{\cdot}000$ m $\quad \frac{1,000}{450} \times 1{\cdot}000$ m $\quad \frac{1,000}{400} \times 1{\cdot}000$ m

= 2·00 m $\quad$ 2·22 m $\quad$ = 2·50 mm

By adding 20% to allow for the end joist the totals are:

2·40 m $\quad$ 2·66 m $\quad$ 3·00 m

Therefore the approximate formula is to divide 1·00 m by the stated spacing, multiply by 1·00 and add 20%.

This formula may also be used for determining the amount of nails required per m² for tongued grooved boarding and for plugs in open-spaced grounds.

Example 12

19 mm white pine dressed, tongued and grooved boarding in 162 mm clear widths to flat roof.

Material

Cost of 19 mm tongued and grooved boarding per m²	272p
Unload and stack	3p
Brads – assume joists at 400 mm centres	
Length of joist per m² = $\frac{1,000}{400} \times 1,000$ = 2·50 mm	
add 20% = 0·50 mm	
3·00 mm	
50 mm brads with two brads per joist	
∴ Number of brads = (3·00 ÷ 0·162) x 2 = 37	
= $\frac{37}{436}$ kgs @ £19·40 per 50 kg	3p say
	278p
Waste 5%	14p
C/fd	292p

B/fd 292p

Labour

Tradesman fixes 1·25 m^2 tongued and grooved boarding per hour

$\therefore$ Cost per $m^2 = \dfrac{220p}{1{\cdot}25}$ 176p

468p

Profit and Oncost 20% 94p

562p

Rate per m^2 £5·62

Example 13

White pine firrings 50 mm wide and average 50 mm deep.

50 mm firrings are taken out of 50 x 100 mm white pine; i.e. 100 m of 50 x 100 mm gives 200 m of firrings. 100 mm of white pine ripped to give 200 m of firrings costs £57·60 plus $\frac{1}{3}$ hour labourer unloading @ 190p = £58·23 5,823p

Nails – 1 per 400 mm = $\dfrac{200{,}000}{400} = 500$

500 nails average 75 mm long = $\dfrac{500}{152}$ kg @ £18·00 for 50 kg 119p

5,942p

Waste 5% 297p

Labour

Tradesman fixes 20·00 m firring per hour

$\therefore$ for 200·00 m tradesman takes 10 hours @ 220p 2,200p

8,439p

Profit and Oncost 20% 1,688p

10,127p

Rate per m = 10,127 ÷ 200 51p

Example 14

38 x 150 mm white pine solid strutting between 50 x 200 mm joists at 450 mm centres.

Material

38 x 150 mm white pine, per 100 m £65·18	
For NET timber there will be no strut at thickness of joist, say 50 mm;	
∴ at 450 mm centres allow $\frac{8}{9}$ x £65·18	5,794p
Unload and stack: $\frac{1}{3}$ hour labourer @ 190p	63p
Nails: with joists at 450 mm centres, 4, 50 mm nails will be required every 450 mm length,	
∴ nails required per 100 m = $\frac{100 \times 4}{0{\cdot}45}$ = 890	
and at 348 per kg = $\frac{890}{348}$ = 2·6 kg @ £18·75 per 50 kg	98p
	5,955p
Waste 5%	298p
	6,253p

Labour

Tradesman fixes 8 m per hour	
∴ Cost per 100 m = 220p x $\frac{100}{8}$	2,750p
	9,003p
Profit and Oncost 20%	1,801p
Cost per 100 m	10,804p
Rate per m	£1·08

Some estimators may ignore the allowance for the thickness of joists as shown above, giving a rate of £1·17 per m.

Example 15

38 x 50 mm white pine herring bone strutting between 50 x 200 mm joists at 450 mm centres.

Material

As joists are at 450 mm centres the space between is 400 mm for 50 mm joists. The length required for each strut is 450 mm and 2 per space are required. That is, 900 mm of timber is required every 450 mm length. Therefore 100 m of timber will provide 50 m of strutting.

38 x 50 mm white pine per 100 m £21·72	2,172p
Unload and stack: $\frac{1}{6}$ hour labourer @ 190p	32p

Nails: with joists at 450 mm centres 8 No. 50 mm (2″) nails will be required every 450 mm length (2 nails per end),

$$\therefore \text{ Nails required per 50 m} = \frac{50 \times 8}{0{\cdot}45} = 890$$

and at 348 per kg = $\frac{890}{348}$ = 2·6 kg @ £18·75 per 50 kg	98p
	2,302p
Waste 5%	115p
	2,417p

Labour

Tradesman fixes 5 m per hour (measured over joists)

$\therefore$ Cost per 50 m = 220p x $\frac{50}{5}$	2,200p
	4,617p
Profit and Oncost 20%	923p
Cost per 50 m (measured over joists)	5,540p
Rate per m	£1·11

Example 16

50 x 75 mm shaped sprockets 500 mm long.

Material

50 x 75 mm white pine per 100 m £43·43	4,343p
Unload and stack: $\frac{1}{3}$ hour labourer @ 190p	63p
Nails: the quantity of timber will provide 200 sprockets and with 4, 75 mm nails per sprocket 800 nails are required.	
∴ at 152 per kg = $\frac{800}{152}$ = 5·3 kg @ £18·00 per 50 kg	191p
	4,597p
Waste 5%	230p
	4,827p

Labour

Tradesman fixes 20 per hour	
∴ Cost per 200 sprockets = 220p x $\frac{200}{20}$	2,200p
	7,027p
Profit and Oncost 20%	1,405p
Cost of 200 sprockets	8,422p
Rate per sprocket	42p

Example 17(a)

19 x 38 mm red pine open spaced grounds set at 400 mm centres plugged to brick walls at 500 mm centres.

Materials

19 x 38 mm red pine grounds cost £11·45 per 100 m	1,145p
Unload and stack: $\frac{1}{6}$ hour labourer @ 190p	32p
	1,177p

1 m² requires 3·00 m of material (*see* pp. 245–6) plus plugs, about 8 No. 50 mm long, which equals 400 mm; total material required for 1 m² is 3·40 m @ £11·77 per 100 m	40p
C/fd	40p

	B/fd	40p
Nails, nominal, say		2p
		42p
Waste 5%		2p
		44p

Labour
Tradesman fixes 3·50 m of grounds per hour including plugging;

$\therefore$ Cost per m² = $220p \times \frac{3{\cdot}40}{3{\cdot}50}$ — 214p

	258p
Profit and Oncost 20%	52p
	310p
Rate per m²	£3·10

Example 17(b)

15 x 38 white pine ground plugged to brickwork at 450 mm centres.

Material
15 x 38 mm softwood sawn ground costs per 100 m £9·40 — 940p
Plugs at 450 m centres and assuming 4 m lengths

$\therefore$ Number of plugs per 4 m length = $\frac{4{,}000}{450}$ = 9 plus one at end = 10

$\therefore$ Number required for 100 m = $\frac{100}{4} \times 10$ = 250 plugs

250 plugs 75 mm long = 18·75 say 19 m	
19 m of 15 x 38 mm ground @ £9·40 per 100 m	179p
Unload and stack grounds ¼ hour labourer @ 190p	48p
Nails for grounds two per plug = 500	
C/fd	1,167p

B/fd	1,167p
50 mm wire nails = $\frac{500}{348}$ = 1·44 kg @ £18·75 per 50 kg	54p
	1,221p
Waste 5%	61p

Labour

Tradesman fixing 100 m of ground including plugs and plugging	
@ 4·50 m per hour = 220p x $\frac{100}{4 \cdot 50}$	4,889p
(output higher than for open-spaced grounds due to plumbing being required on open-spaced grounds in addition to plugging and aligning)	6,171p
Profit and Oncost 20%	1,234p
Cost of 100 m	7,405p
Rate per m	74p

Example 18

19 x 38 mm white pine branders fixed to ceiling joists at 300 mm centres.

Materials

19 x 38 mm white pine grounds cost £8·71 per 100 m	871p
Unload and stack a.b.	32p
	903p

1 m² material required $\left(\frac{1{,}000}{300} \times 1{\cdot}000 \text{ m}\right)$ plus 20% = 4·00 m	
Material required for 1 m² is 4·00 m @ £9·03 per 100 m	36p
Nails, nominal, say	2p
C/fd	38p

	B/fd	38p
Waste 5%		2p
		40p

Labour

Tradesman fixes 8·00 m of brander per hour

$\therefore$ Cost per m² = 220p × $\frac{4 \cdot 00}{8 \cdot 00}$	110p
	150p
Profit and Oncost 20%	30p
	180p
Rate per m²	£1·80

Example 19

Cut do. to angle including 19 x 38 mm angle brander.

1 m² requires 4·00 m of brander costing 180p;

$\therefore$ 1 m of 19 x 38 mm brander (ignoring angle cut) = $\frac{180}{4}$	45p
Rate per m	45p

Example 20

19 x 38 mm white pine bracketing at 300 mm centres encasing steel beams.

Material

Cost of material as previous example is £9·03 per 100 m

Material required per m² = $\left(\frac{1,000}{300} \times 1 \cdot 000\right)$ plus 20% (*see* formula on page 246) = 4·00 m

$\therefore$ 4·00 m @ £9·03 per 100 m		36p
Nails, nominal		2p
		38p
Waste 5%		2p
	C/fd	40p

	B/fd 40p
Labour	
Tradesman fixes 6 m of bracketing per hour,	
$\therefore$ Cost per m² = 220p x $\frac{4\cdot00}{6\cdot00}$	147p
	187p
Profit and Oncost 20%	37p
	224p
Rate per m²	£2.24

Example 21

38 x 50 mm white pine blockings to web of 310 x 250 mm universal beam.

Note: The Standing Joint Committee for S.M.M. have ruled that blocking pieces to web of beams to receive bracketing shall be numbered.

Material	
38 x 50 mm white pine costs £21·72 per 100 m	2,172p
Unload and stack: $\frac{1}{3}$ hour labourer @ 190p	63p
	2,235p
One blocking piece, allow, say, 300 mm long, costs (including waste) $\frac{2{,}235\text{p}}{100}$ x 0·300	7p
Labour	
Tradesman will cut and fix 18 per hour,	
$\therefore$ Cost = $\frac{220\text{p}}{18\text{p}}$	12p
	19p
Profit and Oncost 20%	4p
	23p
Rate each	23p

Example 22

50 x 100 mm white pine in stud partitions.

Note: Timber in partitions may be priced by the cube metre or linear metre. In this case we shall price by linear metre.

Material

50 x 100 mm white pine, per 100 m £53·62	5,362p
Unload and stack: ¼ hour labourer @ 190p	48p
Nails (75 mm): 3 kg @ £18·00 per 50 kg	108p
	5,518p
Waste 5%	276p
	5,794p

Labour

Tradesman fixes 8·50 m per hour (or 0·043 m^3 per hour),

$\therefore$ Cost per 100 m = $220\text{p} \times \dfrac{100}{8{\cdot}5}$	2,588p
	8,382p
Profit and Oncost 20%	1,676p
Cost per 100 m	10,058p
Rate per m	£1·01

13 Woodwork (Joinery)

ROOF JOINER

Example 1

21 x 195 mm red pine fascia board.

Material	
21 x 195 mm red pine per 100 m £79·36	7,936p
Unload and stack: ¼ hour labourer @ 190p	48p
Nails (50 mm): with ceiling joists at 400 mm centres = about 502;	
∴ 1·5 kg nails @ £18·75 per 50 kg	56p
	8,040p
Waste 5%	402p
	8,442p
Labour	
Tradesman fixes 7 m per hour,	
∴ Cost per 100 m = 220p x $\frac{100}{7}$	3,143p
	11,585p
Profit and Oncost 20%	2,317p
Cost per 100 m	13,902p
Rate per m	£1·39

Note: If mitres are required they are priced thus.

Example 2

Tradesman takes 3 minutes per mitre = 20 per hour,

$\therefore$ Cost = $\frac{220p}{20}$	11p
Profit and Oncost 20%	2p
	13p
Rate	13p

Example 3

12 mm red pine tongued and grooved and V-joined soffit board in 75 mm clear widths to soffit 300 mm wide.

Note: Clear widths are finished widths exclusive of tongues.

Material	
Cost of 12 mm tongued and grooved per m²	
£3·22	322p
Unload and stack	3p
	325p
Cost of 1 m 300 mm wide (4 boards)	
$\left(\frac{325p}{1{,}000} \times 300\right)$	98p
30 mm brads 1¼″ nominal	2p
	100p
Waste 5%	5p
	105p
Labour	
Tradesman fixes 3 m per hour,	
$\therefore$ Cost per m = $\frac{220p}{3}$	73p
	178p
Profit and Oncost 20%	36p
	214p
Rate per m	£2·14

Example 4

6 mm asbestos wood soffit 240 mm wide.

Material	
6 mm asbestos sheet 2,440 x 1,220 mm cost per m² £1·79	179p
Unload and stack	3p
	182p
Cost of 1 m 240 mm wide 52p x $\frac{240}{1,000}$	44p
Nails, 25mm nominal	2p
	46p
Waste 7½% for odd size of sheet	3p
	49p
Labour	
Tradesman will rip and fix 4·50 m per hour,	
∴ Cost per m = $\frac{220p}{4 \cdot 50}$	49p
	98p
Profit and Oncost 20%	20p
	118p
Rate per m	£1·18

Example 5

19 mm white pine tongued and grooved flooring in 113 mm clear widths double nailed to each joist with 65 mm brads.

Material	
Cost of 19 mm tongued and grooved flooring per m² £2·79	279p
Unload and stack	3p
C/fd	282p

B/fd	282p
Brads—assume joists at 450 mm centres (*See* Chapter 12, under diagram illustrating open spaced grounds) length of joists per m² = 2·660 m;	
∴ Number of brads = (2·660 ÷ 0·113) x 2 = 47	
$\frac{47}{229} = \frac{1}{5}$ kg @ £18·60 per 50 kg	8p
	290p
Waste 5%	15p
	305p
Labour	
Tradesman fixes 1·7 m² per hour,	
∴ Cost per m² = $\frac{220p}{1·7}$	129p
	434p
Profit and Oncost 20%	87p
	521p
Rate per m²	£5·21

Example 6

22 mm hardwood tongued and grooved flooring in 63 mm clear widths double nailed to each joist with 65 mm brads.

Material	
Cost of 22 mm hardwood tongued and grooved flooring per m²	880p
Unload and stack	5p
Brads: assume joists at 400 mm centres;	
∴ Length of joist = 3·00 m	
∴ Number of brads = (3·000 ÷ 0·063) x 2 = 96	
$\frac{96}{229}$ = 0·42 kg @ £18·60 per 50 kg	16p
	901p
Waste 5%	45p
C/fd	946p

	B/fd 946p
Labour	
Tradesman fixes 0·90 m² per hour,	
∴ Cost per m² = 220p ÷ 0·90	244p
	1,190p
Profit and Oncost 20%	238p
	1,428p
Rate per m²	£14·28

FIXING JOINERY FINISHINGS

Labour Constants for fixing joinery finishings in softwood

Type and width of member	Tradesman hours per 100 m
45 mm architrave, facings, belting, stops	11·50
70 mm do.	12·30
95 mm do.	13·20
120 mm do.	19·70
145 mm do.	20·50
95 mm skirtings	16·40
120 mm do.	17·25
145 mm do.	18·10
145 mm built-up base (2 members)	21·40
215 mm do. 2 do.	24·60
215 mm do. 3 do.	26·25
Sawn ground including plugging to brickwork	4·50 m per hour

Note: (1) The widths of dressed softwood as shown above are standard merchantable finished widths as agreed by the Timber Traders' Association. The thicknesses will be 12, 21, 33, and 45 mm as required. Standard dressed mouldings are also supplied. The sizes stated for dressed softwood are therefore merchantable finished sizes and comply with the Trade Descriptions Act. Softwood, however, may still be purchased in sawn sizes (e.g. 50 x 75 mm sawn would be 45 x 70 mm when dressed) and dressed or ripped down and dressed by contractor as required. The waste incurred in such conversion (10 to 15%) would be allowed for and added to the cost of the member and kept separate from the waste incurred in fixing the finished timber.

(2) Hardwood is generally sold as sawn timber in thicknesses ranging from 25 to 100 mm, and in widths averaging from 150 to 225 mm, dependent on the type of hardwood. Greater widths may also be obtained at additional cost. Hardwood is also sold in solid logs. In both cases it is sold by the cube metre. The cost of hardwood would include the cost of sawing, dressing, running necessary mouldings, and waste incurred in conversion, as distinct from waste incurred in fixing. There is also a selection of dressed standard finishings such as skirtings, architraves and facings, door stops, etc. sold in units of 100 m.

(3) Increase the labour constants for fixing (and/or making joinery finishings as follows:

Type of timber	
Pitch pine	50%
General hardwood, e.g. makoré, agba, limba, meranti	50%
Japanese oak	75%
English oak	100%
Teak	100 to 150%
Mahogany	100 to 150%

Fitting and hanging doors including fixing butts:

Type	Tradesman hours
Standard flush doors	$1\frac{1}{4}$ to $1\frac{1}{2}$
50 mm framed, ledged and braced doors	2 to $2\frac{1}{2}$
38 to 50 mm panelled doors	$1\frac{1}{2}$ to $2\frac{1}{2}$

The time will vary according to size and quality of door.

Example 7

12 x 95 mm softwood skirting rounded one arris and fixed to softwood ground (measured separately)

Material

12 x 95 mm softwood skirting costs per 100 m £39·81	3,981p
Unload and stack skirting labourer ¼ hour @ 190p	48p
Nails for skirting when nailed at 2 per 500 mm centres = 400	
50 mm oval brads = $\frac{400}{348}$ = 1·15 kg @ £19·40 per 50 kg	45p
	4,074p
Waste 5%	204p
Labour	
Tradesman fixing 100 m of skirting takes 16·40 hours @ 220p	3,608p
	7,886p
Profit and Oncost 20%	1,577p
Cost of 100 m	9,463p
Rate per m	95p

Note: For fixing a 16 x 95 mm hardwood (say Makore) skirting the tradesmans time would be 16·40 + 50% = 24·60 hours.

Example 8

44 mm standard flush door size 815 x 1,980 mm faced both sides with 5 mm makoré plywood and lipped with hardwood on two edges (to BS 459, Part 2).

Material

44 mm door costs	1,860p
Take delivery and stack 3 minutes labourer @ 190p	10p
Labour	
Hanging door including fixing butts 1½ hours tradesman @ 220p	330p
C/fd	2,200p

	B/fd	2,200p
Profit and Oncost 20%		440p
		2,640p
Rate		£26·40

Note: On manufactured articles such as doors the contractor may wish to reduce profit margin on material only, in order to obtain a more competitive rate, e.g. 10%. For example:

Material	
44 mm door costs, including unloading and stacking, as before	1,870p
Profit and Oncost on material 10%	187p
	2,057p
Labour	
Hanging door, as before	330p
Profit and Oncost on labour 20% on 330p	66p
	2,453p
Rate	£24·53

Note: No waste is allowed on standard doors but the estimator may wish to allow, say, 1% for damage in handling.

Example 9

45 x 120 mm softwood plain door lining with tongued angles fixed to billgates (measured separately).

Material	
45 x 120 mm door lining costs per 100 m £75·71	7,571p
Unload and stack: $\frac{1}{3}$ hour labourer @ 190p	63p
Nails 90 mm: the average length of a set of linings for a single door is 5·00 m and each leg is double-nailed at three points, which requires 12 for fixing, and 6 are required for making, totalling 18 nails per set;	
C/fd	7,634p

	B/fd 7,634p
Total nails required per 100 m = $18 \times \frac{100}{5}$	
$= 360; \frac{360}{108} = 3{\cdot}33$ kg @ £17·80 per 50 kg	119p
	7,753p
Waste 5%	388p
	8,141p
Labour	
Tradesman makes and fixes 2·50 m per hour	
$\therefore$ Per 100 m = $220\text{p} \times \frac{100}{2{\cdot}50}$	8,800p
	16,941p
Profit and Oncost 20%	3,388p
Cost per 100 m	20,329p
Rate per m	£2·03

Example 10

12 x 70 mm softwood door facing rounded on two arrises.

Material	
12 x 70 mm door facing costs per 100 m £20·07	2,007p
Unload and stack: ¼ hour labourer @ 190p	48p
50 mm brads using 4 brads per m, staggered = 400 brads	
$\frac{400}{436}$ = say 1 kg @ £19·40 per 50 kg	39p
	2,094p
Waste 5%	105p
	C/fd 2,199p

B/fd	2,199p
Labour	
Tradesman fixing 100 m takes 12·30 hours @ £2·20	2,706p
	4,905p
Profit and Oncost 20%	981p
Cost per 100 m	5,886p
Rate per m	59p

Example 11

21 x 45 mm softwood door stop.

Material	
21 x 45 mm door stop costs per 100 m £19·03	1,903p
Unload and stack: ¼ hour labourer @ 190p	48p
50 mm brads as previous example	39p
	1,990p
Waste 5%	100p
	2,090p
Labour	
Tradesman fixing 100 m takes 11·50 hours @ 220p	2,530p
	4,620p
Profit and Oncost 20%	924p
Cost per 100 m	5,544p
Rate per m	55p

Example 12

16 x 70 mm makoré door facing rounded on two arrises.

Material	
16 x 70 mm door facing costs per 100 m £66·45	6,445p
C/fd	6,445p

B/fd	6,445p
Unload and stack: ¼ hour labourer @ 190p	48p
50 mm brads as Example 10	39p
	6,532p
Waste 5%	327p
	6,859p
Labour	
Tradesman fixing 100 m takes 12·20 hours plus 50% = 18·45 hours @ 220p	4,059p
	10,918p
Profit and Oncost 20%	2,184p
Cost of 100 m	13,102p
Rate per m	£1·31

Example 13

Standard ready-made softwood casement window complete with frames to comply with BS 644 Part 1, type 6CVC35 size 1,800 x 1,050 mm overall complete with steel easy-clean hinges and ironmongery.

Material	
Cost per window delivered site £31·95	3,195p
Unload and stack: 3 minutes labourer @ 190p	10p
25 x 100 mm sawn billgates 225 mm long supplied to bricklayer for building in, 4 @ 7p	28p
Nails, nominal	2p
Labour	
Tradesman uplifting, setting and fixing ¾ hour @ 220p	165p
	3,400p
Profit and Oncost 20%	680p
	4,080p
Rate each	£40·80

Note: (1) The profit may be restricted in a similar manner to standard doors.

(2) The average time for unloading and stacking is about 3 to 5 minutes per window, i.e. if 2 labourers are used the time is $1\frac{1}{2}$ minutes each.

(3) In some areas this type of standard window is set in position by bricklayer and nailed by joiner if required. In other areas the windows are set and fixed by joiners after the brickwork has been completed.

(4) For setting windows the time required by tradesmen ranges from $\frac{1}{2}$ hour for small windows to 1 hour for large windows. When set by bricklayers the time is about $\frac{1}{4}$ hour for bricklayer and labourer for small windows to $\frac{1}{2}$ hour for bricklayer and labourer for large windows.

(5) For plugging and screwing stiles of windows allow 4·50 m per hour plus 4 screws per m.

Example 14

33 x 170 mm softwood window-board, rounded one arris, tongued and grooved to sill and plugged to brick.

Materials

33 x 170 mm window-board costs per 100 m £91·41	9,141p
25 x 100 mm red pine plugs each about 100 mm long at 450 mm centres; the average is about 2·5 per m to suit various lengths of window-board, i.e. 250 per 100 m;	
∴ 100 mm x 250 = 25 m of 25 x 100 mm red pine @ £37·86 per 100 m	947p
$2\frac{1}{2}$" (63 mm) brads at 2 nails per plug = 500	
$\frac{500}{229}$ = 2·14 kg @ £18·60 per 50 kg	80p
	10,168p
Waste 5%	508p
	10,676p

Labour

Tradesman plugging 3·00 m per hour = 220p $\times \frac{100}{3 \cdot 00}$	7,333p
C/fd	18,009p

B/fd	18,009p
Tradesman fitting window-board 3·50 m per hour = 220p × $\frac{100}{3 \cdot 50}$	6,286p
	24,295p
Profit and Oncost 20%	4,859p
Cost per 100 m	29,154p
Rate per m	£2·92

Example 15

Returned, notched end of do.

Tradesman will mark out, cut rebate, and form return in about 20 minutes.

Tradesman $\frac{1}{3}$ hour @ 220p	73p
Profit and Oncost 20%	15p
	88p
Rate	88p

Example 16

Standard ready-made softwood double hung sash window complete with solid frame to comply with BS 644 Part 2, type 35S57N size 1,054 × 1,712 mm complete.

Material	
Cost per window delivered site £60·00	6,000p
Unload and stack: 5 minutes labourer @ 190p	16p
Wedges 4 @ 3p	12p
Labour	
Tradesman uplifting, setting, and wedging $\frac{3}{4}$ hour @ 220p	165p
	6,193p
Profit and Oncost 20%	1,239p
	7,432p
Rate each	£74·32

Example 17

Fit and hang sash 'X' type spring balance complete:

Material	
Cost of one pair of spring balances including screws @ £8·00	800p
Labour	
Tradesman taking out sash, fitting balance, and sash, ¾ hour @ 220p	165p
	965p
Profit and Oncost 20%	193p
	1,158p
Rate each	£11·58

Note: (1) The average time for unloading and stacking is about 4 minutes per window.

(2) For setting windows the time required by tradesman ranges from ½ hour for small windows to 2 hours for large windows.

(3) For taking out and fitting and hanging a sash with sash cord or spiral balances the average time is about ¾ hour.

(4) The length of sash cord required is normally about twice the height of the sash.

FITTING JOINERY COMPONENTS

The manufacture of joinery components is a highly specialized industry and standard ready-made items such as windows, doors, stairs, kitchen fitments, etc. may be purchased much more economically than making them in a joiner's shop.

However, there are numerous cases where such items are not to standard sizes and are in relatively small quantities; it may be more convenient, and in some cases more economic, to make the items in the joiner's shop.

Window Frames

The following items are for a shop-made casement window:

The number of mortice and tenon joints must be assessed

as an average per linear metre as follows:

Assume one window size 1·00 x 1·00 m, and one size 2·00 x 1·25 m.

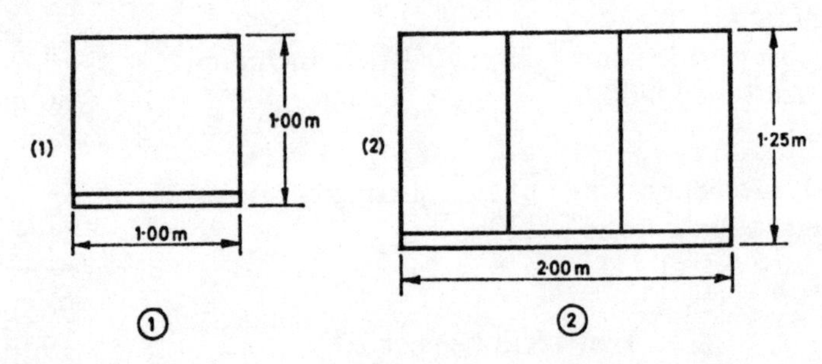

At window (1) there are 4 mortice and tenon joints including those into sill frame, i.e. 4 mortice and tenon joints per 4·00 m of frame (including sill), which is 1 mortice and tenon joint per 1·00 lin.m.

At window (2) there are 4 mortice and tenon joints (excluding joints to mullion frames) per 6·50 m of frame (including sill), which is 1 joint per 1·625 m.

A reasonable average between varying sizes of windows would be 1 mortice and tenon joint per 1·25 lin.m.

For mullion frames a reasonable average would be 1 mortice and tenon joint per 0·60 lin.m.

For transom frames a reasonable average would be 3 mortice and tenon joints per linear metre.

Example 18

70 x 95 mm softwood (red pine) frame, rebated, grooved and splayed.

Material

70 x 95 mm red pine frame (machined as described) costs per 100 m £114·50	11,450p
Unload and stack: $\frac{1}{3}$ hour labourer @ 190p	63p
	11,513p
∴ Cost per m	115p
Nails, wedges, etc., nominal	3p
White lead per joint	3p
C/fd	121p

	B/fd 121p
Waste 5%	6p
	127p

Labour
Tradesman frames up and fits in position 2·00 m per hour

∴ 1 m costs 220p x $\frac{1}{2{\cdot}00}$	110p
Mortice and tenon joints of frames: assume machine costs £3·20 per hour including operator; ∴ Marking out and cutting at 10 minutes per joint = 320p ÷ 6 = 53p per joint at 1 joint per 1·25 m = 53p x $\frac{4}{5}$	42p
	279p
Profit and Oncost 20%	56p
	335p
Rate per m	£3·35

Example 19

70 x 145 mm softwood (red pine) sill frame, double-sunk, weathered, twice throated, and grooved.

Material

70 x 145 mm red pine sill frame (machined as described), cost per 100 m £176·00	17,600p
Unload and stack: $\frac{1}{2}$ hour labourer @ 190p	95p
	17,695p
∴ Cost per m	177p
Nails, wedges, etc., nominal	3p
White lead paint per joint	3p
	C/fd 183p

	B/fd 183p
Waste 5%	9p
	192p

Labour
Tradesman frames up and fits in position
2·00 m per hour;

∴ 1 m costs = 220p x $\frac{1}{2\cdot00}$	110p
Mortice and tenon joints of frame as Example 17	42p
	344p
Profit and Oncost 20%	69p
	413p
Rate per m	£4·13

Example 20

45 mm thick softwood casement splayed and prepared for glazing.

Consider a casement of size 450 x 900 mm;
∴ Quantities of timber required
= 2/0·45 = 0·90
= 2/0·90 = 1·80
2·70

Materials

2·70 m of 45 x 45 mm splayed and rebated red pine @ 49p per m	132p
Nails, wedges, white lead paint for joints, and unloading, etc., say	25p
	157p
(*Note:* Waste in making joinery varies from 10 to 20%) Waste 15%	24p
	C/fd 181p

	B/fd 181p
Labour	
From records a wood-working machine costs 50p per hour to run; ∴ tradesman @ 220p and machine @ 50p = 270p	
Tradesman and machine, morticing and tenoning to joints = ½ hour @ 270p	135p
Tradesman fabricating casement ½ hour @ 220p	110p
Allow for haulage to site	15p
	441p
Profit and Oncost 20%	88p
Cost of casement size 450 x 900 mm	529p
Rate per m² = 529p ÷ 0·405 m² (area of casement)	£13·06

Example 21

Fit and hang casement

Labour	
Average time for tradesman fitting and hanging casement 40 minutes	
Tradesman fitting and hanging casement = 40 minutes @ 220p	147p
Profit and Oncost 20%	29p
	176p
Rate	£1·76

In order to assess the total cost of a purpose-made window of a stated size a compilation is made of the total cost of frame, sill, mullion (if required) the casement plus fitting and hanging, plus the cost of the hinges.

Ironmongery is measured separately.

Glazing Beads

Example 22

8 x 15 mm softwood rounded glazing beads.

Material	
100 m of standard ready-run glazing bead, per 100 m £9·48	948p
Unload and stack: 5 minutes labourer @ 190p	16p
As glazing beads are normally short lengths, assume 1 pin about every 150 mm, say 7 per m = 700 per 100 m;	
∴ pins about $\frac{1}{3}$ kg @ £21·00 per 50 kg, say	14p
	978p
For short lengths, waste 15%	147p
	1,125p
Labour	
Tradesman will cut and mitre and temporarily fit 6·50 m per hour;	
∴ for 100 m tradesman = 220p x $\frac{100}{6 \cdot 50}$	3,385p
	4,510p
Profit and Oncost 20%	902p
Cost per 100 m	5,412p
Rate per m	54p

Example 23

8 x 15 mm hardwood glazing beads secured with brass cups and screws.

Material	
100 m of hardwood stadard ready-run glazing bead, per 100 m £40·20	4,020p
Unload and stack: $7\frac{1}{2}$ minutes labourer @ 190p	24p
Cups and screws are required about every 150 mm for short lengths	
(1,000 x 100) ÷ 150 = 667	
667 19 mm brass screws @ 180p per gross	833p
667 brass cups @ 45p per gross	208p
C/fd	5,085p

	5,085p
Waste 15%	763p
	5,848p

Labour
Tradesman wil cut and mitre and temporarily fit 2·50 m per hour;

$\therefore$ 100 m tradesman = 220p x $\frac{100}{2{\cdot}50}$	8,800p
	14,648p
Profit and Oncost 20%	2,930p
Cost per 100 m	17,578p
Rate per m	£1·76

Door Frames

Example 24

45 mm (finished thickness) bound softwood three-panel door size 900 x 2,000 mm comprising 110 mm top rail, stiles, and muntin, 245 mm bottom, and 195 mm lock rail, moulded on solid, both sides, the two lower panels of 12·5 mm thick Douglas fir plywood, the upper panel divided into six small panes for glazing with 33 x 45 mm twice-moulded and twice-rebated glazing bars.

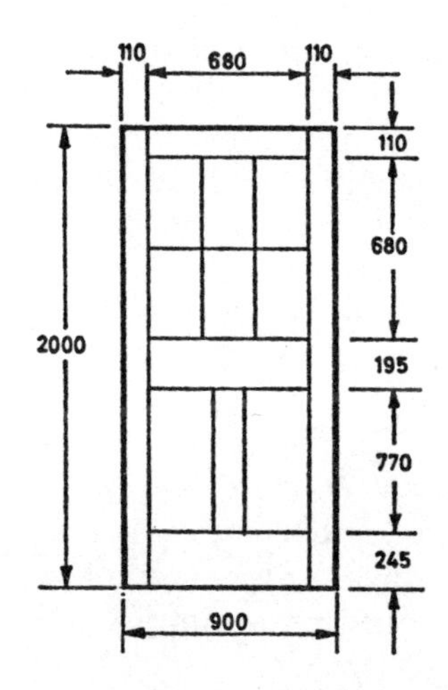

Quantities required

Red pine dressed

45 x 110 mm stiles 2/2·00	4·00	
top rails	0·90	
muntin	1·21	
	6·11 m @ 90p per m	550p
45 x 245 m bottom rail, 0·90 m @ 180p per m		162p
45 x 195 mm lock rail, 0·90 m @ 145p per m		131p
33 x 45 mm glazing bars 3/0·73, 2·19 m @ 34p per m		74p
12·5 mm Douglas fir plywood 2 pieces each size 315 x 820 mm = 0·52 m² @ £3·23 per m²		168p
		1,085p
Waste 15%		163p
		1,248p

Note: Timber is assumed to be purchased as dressed timber; if not, an allowance for dressing must be made.

Labour

Tradesman and machine, morticing, tenoning, and cutting out door = 2 hours @ 274p	548p
Tradesman and machine, moulding and rebating stiles and glazing bars = 2 hours @ 274p	548p
Tradesman fabricating door and jointing 4 hours @ 220p	880p
Tradesman hanging door on site 2 hours @ 220p	440p
Loading up, transporting and unloading at site	36p
	3,700p
Profit and Oncost 20%	740p
	4,440p
Rate	£44·40

Note: The allowance for loading and transporting to site may vary according to circumstances.

Example 25

45 mm (finished thickness) bound makoré single-panel glazed door size 900 x 2,000 mm comprising 110 mm top rail and stiles, 245 mm bottom rail moulded on solid one side and rebated, and divided into small panes for glazing with 33 x 45 mm twice-moulded and twice-rebated glazing bars.

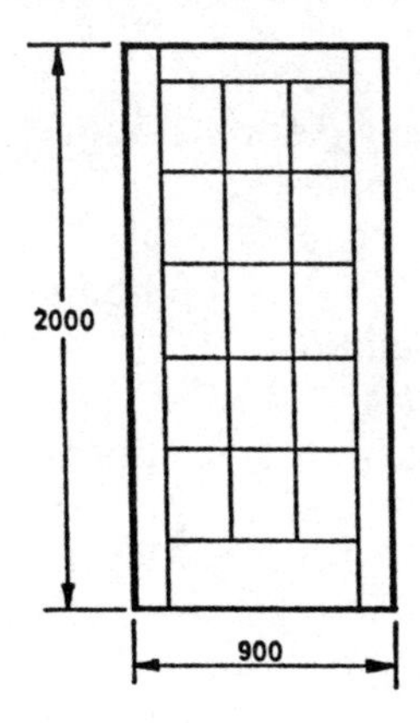

Quantities Required

45 x 110 m stiles	2/2·00 = 4·00		
top rail	= 0·90		
	4·90 m @ 173p per m		848p
45 x 245 mm bottom rail	0·90 m @ 390p per m		351p
			1,199p
33 x 45 mm glazing bars	2/1·70 = 3·40		
	4/0·68 = 2·72		
	6·12 m		
	6·12 m @ 55p per m		337p
		C/fd	1,536p

		B/fd	1,536p
Note: 20% for waste is allowed as timber is cut from square-edged lumber			
Waste 20%			307p
			1,843p
Labour			
Tradesman and machine ripping and dressing timber	1 hour		
Tradesman and machine morticing, tenoning, and cutting our door	3 hours		
Tradesman and machine moulding and rebating stiles and glazing bars	3 hours		
	7 hours @ 274p		1,918p
Tradesman fabricating door and jointing	6 hours		
Tradesman hanging door on site	3 hours		
	9 hours @ 220p		1,980p
Loading up and transporting to site and unloading			54p
			5,795p
	Profit and Oncost 20%		1,159p
			6,954p
	Rate		£69·54

Example 26

70 x 95 mm softwood rebated door frame plugged to brickwork.

Material		
70 x 95 mm red pine rebated frame cost per 100 m £117·32		11,732p
	C/fd	11,732p

B/fd	11,732p
25 x 100 mm sawn red pine plugs; the average length of a set of frames for a single door is about 5·33 m and each set requires 6 plugs 150 mm long.	
Number of plugs required per 100 m =	
$6 \times \frac{100}{5{\cdot}33} = 113$	
113 @ 150 mm = say 17 m of 25 x 100 mm @ £34·30 per 100 m	583p
Unload and stack timber: ½ hour labourer @ 190p	95p
Nails = 18 per set of frames (*see* Example 9)	
∴ Number required per 100 m =	
$18 \times \frac{100}{5{\cdot}33} = 338$	
$\frac{338}{31} = 10{\cdot}9$ kg @ £17·00 per 50 kg	371p
	12,781p
Waste 5%	639p
	13,420p
Labour	
Tradesman makes and fixes 2·00 m per hour including plugging	
∴ Per 100 m = $220\text{p} \times \frac{100}{2{\cdot}00}$	11,000p
	24,420p
Profit and Oncost 20%	4,884p
Cost per 100 m	29,304p
Rate per m	£2·93

Stairs

Example 27

Parana pine steps comprising 28 mm tread, in one width, with rounded nosing and 19 x 19 mm bed mould, 21 mm riser, tongued, grooved and screwed together including 63 x 100 mm sawn white pine rough carriage with 25 x 150 mm rough brackets, glued blockings and wedges complete.

Consider a step 1·00 m long, with tread 295 mm wide and riser (net) 155 mm deep; ∴ Area = 1·00 x (0·295 + 0·155)
= 0·45 m²

Material	
Tread 100 x 295 mm wide and 28 mm thick	210p
Riser 1·00 x 150 mm wide and 21 mm thick	79p
Bed mould 1·00 m 19 x 19 mm	20p
Blockings 2 @ 5p	10p
Rough carriage requires about 300 mm per tread 0·30 m of 63 x 100 sawn white pine @ 70p per m	21p
25 x 150 mm rough bracket 250 mm long @ 44p per m	11p
Wedges, screws, and glue	9p
Unload and stack materials generally	3p
	363p
Waste 10%	36p
	399p
Labour	
Tradesman making, fitting, erecting, 0·20 m² per hour; area is 0·45 m²,	
∴ 220p x $\frac{0{\cdot}45}{0{\cdot}20}$	495p
	894p
Profit and Oncost 20%	179p
Cost per step (area 0·45 m²)	1,073p
Rate per m² 1,073p x $\frac{1}{0{\cdot}45}$	2,384p
	£23·84

Example 28

33 x 295 mm Parana pine outer string, in one width.

Materials	
33 x 295 mm outer string costs per m	235p
Unload and stack	3p
	238p
Waste 10%	24p
	262p
Labour	
Tradesman 3·00 m per hour @ 220p =	
220p ÷ 3	73p
	335p
Profit and Oncost 20%	67p
	402p
Rate per m	£4·02

Suggested average times for work on stairs for tradesman:

In order to assess the total cost of a stair, a compilation is made of the total cost of steps, outer and wall strings, newels, handrails and balusters with all necessary labours plus the cost of associated half or quarter landings.

Fixing wall string	2·00 m per hour
Fixing newels	1·25 m per hour
Fixing handrails	5·00 m per hour

Housing ends of steps with rounded nosing, each	0·40 hours
Housing ends of steps with moulded nosings, each	0·50 hours
Housing ends of string to newel each	0·50 hours
Housing ends of handrail to newel each	0·30 hours

Lining to walls

Example 29

19 mm Makore tongued, grooved and channelled vertical lining

to walls in 90 mm clear widths, fixed with panel pins to softwood open-spaced grounds (measured separately).

Note: This type of timber is sold by suppliers stating widths laid and the extra timber for tongues, rebates etc. is included in the purchase price.

Material	
19 mm Makore lining per m² £12·80	1,280p
Unload and stack say	5p
Panel pins: grounds at 450 mm centres (*see* pp. 245–6); length of ground per m² = 2·66 m;	
∴ Number of panel pins = (2·66 m + 90 mm) x 2 (2 pins to each joint) = 60 ∴ allow	3p
	1,288p
Waste 5%	64p
Labour	
Tradesman fixes 0·60 m² lining per hour @ 220p = 220p ÷ 0·60	367p
	1,719p
Profit and Oncost 20%	344p
	2,063p
Rate per m²	£20·63

Example 30

5 mm Austrian oak veneered plywood lining to walls with plain butt joints, fixed with approved adhesive and panel pins to softwood open-spaced grounds (measured separately).

Note: (1) Grounds are known to be placed horizontally and vertically at 610 mm centres;

(2) Veneered plywood sheets are sold in sizes of 2,440 x 1,220 mm, and 2,135 x 915 mm, and in units of 10 m²;

(3) Waste on wall lining may vary from 5% to 15% dependant on wall heights, etc. We will allow 10% in this example.

Material

5 mm veneered plywood lining costs per 10 m² £63·95	6,395p
Unload and stack 6 minutes labourer @ 190p	19p
	6,414p
Cost per m²	641p
Panel pins at 250 mm centres allow 20 per m²; Adhesive; Plastic wood infilling — nominal	9p
	650p
Waste 10%	65p
Labour	
Tradesman fixes 2·00 m² plywood lining per hour @ 220p = 220p ÷ 2	110p
	825p
Profit and Oncost 20%	165p
	990p
Rate per m²	£9·90

IRONMONGERY

Ironmongery items may be described in a Bill of Quantities, to avoid confusion, as 'supply and fix', which means that the cost of the component (plus fixing, profit, etc.) should be included in the price of the item. The prime cost of each component may be included in the item description. By stating the prime cost of each item the scope of choice remains open to the architect should a new range of ironmongery come on to the market.

The alternative is to allow a Prime Cost Sum for ironmongery in the Bill of Quantities with an item for profit and to bill items as:

'Fix the following ironmongery' (*a*) to softwood;
(*b*) to hardwood.

Note: In accordance with S.M.M. screws are deemed to be included in price of ironmongery.

Suggested average times for tradesman fixing ironmongery to softwood:

	Minutes
3 to 6″ (75 to 150 mm) tower bolts	18
9 to 12″ (225 to 300 mm) tower bolts	30
3 to 6″ (75 to 150 mm) flush bolts	40
9 to 12″ (225 to 300 mm) flush bolts	60
75 mm cupboard lock	35
Rim lock and furniture	60
Mortice lock and furniture	90
Postal knocker	90
Casement stay	18
Casement fastener	12
Finger plate	10
Hat and coat hooks	5

For fitting to hardwood allow 25% increase in time.

Example 31

Supply and fix 100 mm butt hinges to softwood.

Material	
Cost of 1 pair butt hinges @ 32p each	64p
Screws, 16 @ 144p per gross	16p
Waste on screws 10%	2p
	82p
Profit and Oncost 20%	16p
	98p
Rate per pair	98p

Note: (1) The labour for fixing hinges is included in the fixing time for hanging door, or window sash.

(2) Hinges are not normally included in the Prime Cost Sum for ironmongery. They are normally billed as supply and fix.

Example 32

Supply and fix 5″ (127 mm) mortice lock with furniture both sides (Prime Cost £12·50)

Material	
Lock, furniture, and screws, cost	1,250p
Labour	
1½ hours tradesman @ 220p	330p
	1,580p
Profit and Oncost 20%	316p
	1,896p
Rate	£18·96

Example 33

Fit: Rim lock and furniture complete.

Materials – Nill

Labour	
1 hour tradesman @ 220p	220p
Profit and Oncost 20%	44p
	264p
Rate for fixing	£2·64

14 Structural Steelwork

Structural steelwork is normally carried out by specialist sub-contractors, and quotations for this type of work are normally obtained from such specialists.

For small contracts where there may be a relatively small amount of, fabricated steelwork, the contractor may wish to purchase and erect the steelwork without using specialists.

The following are suggested average times for unloading, hoisting and setting steelwork.

	Erector and labourer
Unloading, hoisting (average 8·00 m high) and setting per tonne (1,000 kg)	
Isolated beams not bolted (unfabricated)	20·0 hours
Beams in fabricated steelwork	40·0 hours
Stanchions in fabricated steelwork	60·0 hours
Roof trusses	80·0 hours

Note: (1) No allowance is normally made for waste.

(2) Lead money of up to 30p per hour may be allowed over labourer's rate for erector, but is not included in the Working Rules.

FABRICATED STEELWORK

Example 1

Universal beam in 254 x 102 mm section exceeding 20 kg but not exceeding 50 kg per lin.m.

Material	
254 x 102 mm section @ £300·00 per tonne	30,000p
Labour	
Unload, hoist and set 40 hours erector @ 220p	8,800p
40 hours labourer @ 190p	7,600p
C/fd	46,400p

	B/fd	46,400p
Profit and Oncost 10%		4,640p
		51,040p
Rate per tonne		£510·40

15 Plumbing Installations

Note: Whilst cast iron is still being used in plumbing installations it is not now used so extensively as in the past. The sizes are still being quoted in Imperial units showing the *nominal* metric equivalent.

CAST IRON RAINWATER PIPES AND GUTTERS

Example 1

5″ (125 mm) diameter cast iron half-round eaves gutter of $\frac{1}{8}$″ (3·2 mm) metal secured to timber fascia with galvanized hooks at 2′ 0″ (610 mm) centres, jointed with red lead putty and gutter bolts.

Cost per 6′ 0″ (1·829 m) length	300p
Take delivery, say	3p
Gutter hooks £2·96 per doz.	
∴ 3 per length	74p
6 screws @ 90p per gross	4p
Joint	
1 gutter bolt and nut @ £1·80 per gross	2p
Jointing compound (red lead), say	5p
	388p
Waste 5%	19p
	407p
Labour	
$\frac{3}{4}$ hour plumber per length @ 230p	173p
	580p
Profit and Oncost 20%	116p
6′ 0″ length costs	696p
% Rate per m = 696p ÷ 1·829	£3·81

Note: External and internal angles, outlets and stop ends are measured as extra over gutters.

The fitting is therefore priced full value and a commensurate length of pipe is deducted.

Deductions to be made for the effective lengths of the various fittings.

Reference BS 460

(1) Double socket nozzle piece, 9½″ (242 mm)
(2) Stop end outlet with socket, 6¾″ (171·5 mm)
(3) Stop end outlet with spigot, 5″ (127 mm)
(4) Square angle 4″ + diameter (say 4½″), 8½″ (101·6 + 114·3 mm) = (216 mm)
(5) Stop end for spigot or socket, Nil

Plumber's time to fit (1) to (4) is 25 minutes per fitting.
Plumber's time to fit (5) is 15 minutes per fitting.

Example 2

Extra over 5″ (125 mm) eaves gutter for double socket nozzle piece.

Cost of nozzle piece	300p
Take delivery	2p
Joint (*Note:* Two joints are required)	
2 gutter bolts and nuts and 1 extra bracket	29p
Jointing compound (red lead), say	10p
	341p
Waste 5%	17p
	358p
Labour	
25 minutes plumber @ 230p	96p
	454p
Profit and Oncost 20%	91p
	545p
Deduct 0·24 m of gutter @ 381p per m	91p
	454p
Rate	£4·54

Note: One joint only is required for the other fittings shown above.

Example 3

3″ (75 mm) diameter cast iron rainwater pipe of $\frac{1}{8}$″ (3·2 mm) metal jointed with rope yarn and red lead putty and secured to brick walls with malleable iron holderbats at 6·0 ft (1·829 m) centres.

Cost per 6′ 0″ (1·829 m) length	416p
Take delivery	5p
One holderbat	75p
Joint red lead say	10p
rope yarn	2p
	508p
Waste 5%	25p
Labour	
Plumbing pipe, making joint and fixing holderbat	
55 minutes plumber @ 230p	211p
	744p
Profit and Oncost 20%	149p
6′ 0″ length costs	893p
Rate per m = 893p ÷ 1·829	£4·88

Note: Bends, offset bends, branches, and shoes are measured as extra over pipes.

Deductions to be made for the effective lengths of various fittings as ascertained from BS 460 : 1964, i.e. from faucet to faucet as shown.

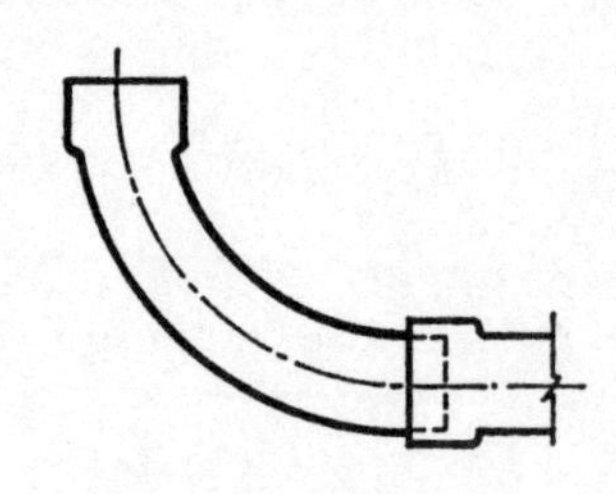

Bends

	2″ (50·4 mm)	$2\frac{1}{2}$″ (63·5 mm)	3″ (76·2 mm)	4″ (101·6 mm)
$92\frac{1}{2}$°	$7\frac{3}{4}$″ (197 mm)	$8\frac{1}{2}$″ (216 mm)	9″ (229 mm)	10″ (254 mm)
$112\frac{1}{2}$°	$6\frac{1}{2}$″ (165 mm)	7″ (178 mm)	$7\frac{1}{4}$″ (184 mm)	8″ (203 mm)
135°	$5\frac{1}{2}$″ (140 mm)	6″ (152 mm)	$6\frac{1}{4}$″ (159 mm)	$6\frac{1}{2}$″ (165 mm)

Single Branches $92\frac{1}{2}$°

	2″ (50·4 mm)		$2\frac{1}{2}$″ (63·5 mm)		3″ (76·2 mm)		4″ (101·6 mm)	
Length	$7\frac{1}{2}$″	191	$8\frac{1}{4}$″	210	$8\frac{3}{8}$″	213	$8\frac{5}{8}$″	219
Branch	$5\frac{1}{8}$″	130	$5\frac{5}{8}$″	143	$5\frac{7}{8}$″	149	$6\frac{3}{8}$″	167
Total Girth	$12\frac{5}{8}$″	321 mm	$13\frac{7}{8}$″	353 mm	$14\frac{1}{4}$″	362 mm	15″	386 mm

Offsets

Projection	2″ (50·4 mm)	$2\frac{1}{2}$″ (63·5)	3″ (76·2 mm)	4″ (101·6 mm)
3″ (76·2)	$10\frac{1}{4}$″ (260 mm)	$10\frac{1}{2}$″ (267 mm)	$10\frac{7}{8}$″ (276 mm)	$11\frac{5}{8}$″ (295 mm)
$4\frac{1}{2}$″ (114·3)	$12\frac{3}{8}$″ (314 mm)	$12\frac{3}{4}$″ (324 mm)	13″ (330 mm)	$13\frac{3}{4}$″ (349 mm)
6″ (152·4)	$14\frac{1}{2}$″ (368 mm)	$14\frac{7}{8}$″ (378 mm)	$15\frac{1}{8}$″ (384 mm)	$15\frac{3}{4}$″ (400 mm)
9″ (228·6)	$18\frac{3}{4}$″ (476 mm)	$19\frac{1}{8}$″ (486 mm)	$19\frac{3}{8}$″ (492 mm)	20″ (508 mm)
12″ (304·8)	23″ (584 mm)	$23\frac{3}{8}$″ (593 mm)	$23\frac{5}{8}$″ (600 mm)	$24\frac{1}{4}$″ (616 mm)

Rainwater Shoes

2″ (50·4 mm)	$2\frac{1}{2}$″ (63·5 mm)	3″ (76·2 mm)	4″ (101·6 mm)
$5\frac{3}{4}$″ (146 mm)	$6\frac{1}{2}$″ (165 mm)	7″ (178 mm)	$8\frac{1}{8}$″ (206 mm)

Example 4

Extra over 3″ (75 mm) diameter cast iron rainwater pipe for offset bend with 12″ (305 mm) projection

Cost of offset		203p
Take delivery		2p
Joint		
Red lead and rope yarn a.b.		12p
		217p
Waste 5%		11p
	C/fd	228p

	B/fd	228p
Labour		
Making joint and fixing (including cutting pipe if required)		
30 minutes plumber @ 230p		115p
		343p
Profit and Oncost 20%		69p
		412p
Deduct 0·60 m of pipe at 488p		293p
		119p
Rate		£1·19

Example 5

Cast iron rainwater head size 10 x 7 x 7″ (254 x 178 x 178 mm) with outlet for 3″ (75 mm) diameter pipe fixed to brick walls with screws.

Note: Rainwater heads are *measured full value*, and *not* extra over pipe.

Cost of rainwater head	390p
Joint red lead and rope a.b.	12p
2 hardwood plugs, say	2p
2 (75 mm) galvanized screws @ 30p per doz.	5p
	409p
Waste 2½%	10p
	419p
Labour	
Making joint and screwing to wall	
½ hour plumber @ 230p	115p
	534p
Profit and Oncost 20%	107p
	641p
Rate	£6·41

CAST IRON SOIL AND WASTE PIPES

Example 6

$3\frac{1}{2}''$ (90 mm) diameter case iron soil and vent pipe of $\frac{3}{16}''$ (4·8 mm) metal jointed with rope yarn and molten lead and secured to brick walls with malleable iron holderbats at 6·0 ft (1·829 m) centres including cutting and pinning.

Cost per 6′ 0″ (1·829 m) length	710p
Take delivery	5p
One holderbat	75p
Joint bat lead 1·6 kg @ 25p per kg	40p
rope yarn 0·1 kg @ 36p per kg	4p
	834p
Waste 5%	42p
Labour	
Plumbing pipe, making joint and fixing holderbat 65 minutes plumber @ 230p	249p
	1,125p
Profit and Oncost 20%	225p
6′ 0″ length costs	1,350p
Rate per m = 1,350p ÷ 1·829	£7·38

Note: The deductions to be made for the effective lengths of various fittings may be ascertained from BS 416: 1967.

Example 7

Extra over do. for $3\frac{1}{2}''$ (90 mm) plain bends.

Cost of bend	282p
Take delivery	2p
Joint lead and yarn a.b.	44p
	328p
Waste 5%	16p
C/fd	344p

B/fd	344p
Labour	
Making joint, cutting pipe as required and fixing	
50 minutes plumber @ 230p	192p
	536p
Profit and Oncost 20%	107p
	643p
Deduct 0·27 m of pipe @ 738p per m	177p
	466p
Rate	£4·66

Example 8

Extra over do. for 3½″ (90 mm) plain branch.

Cost of branch	396p
Take delivery	2p
Joint (two joints are required for branch)	
Lead and rope yarn for two joints	88p
	486p
Waste 5%	24p
Labour	
Making two joints: cutting pipe as required and fixing	
1½ hours plumber @ 230p	345p
	855p
Profit and Oncost 20%	171p
	1,026p
Deduct 0·37 m of pipe @ 738p per m	273p
	753p
Rate	£7·53

PVC RAINWATER PIPES AND GUTTERS

Example 9

100 mm PVC heavy grade half round eaves gutters, with snap joints, secured to timber with fascia brackets at 610 mm centres.

Cost per 2·00 m length	204p
Take delivery	2p
Fascia brackets 3·3 @ 19p	63p
6 screws @ 144p per gross	6p
	275p
Waste 5%	14p
Labour	
40 minutes plumber per length @ 230p	153p
	442p
Profit and Oncost 20%	88p
2 m length costs	530p
Rate per m = 530p ÷ 2	£2·65

Example 10

Extra over 100 mm eaves gutter for stop end.

Cost of stop end	29p
Take delivery	1p
	30p
Waste $2\frac{1}{2}$%	1p
Labour	
10 minutes plumber @ 230p	38p
	69p
Profit and Oncost 20%	14p
Rate	83p

Note: No deduction is made for extra for stop end as there is no extra length required.

Example 11

68 mm PVC heavy grade rainwater pipe with socket on one end of each length jointed with neoprene 0 ring and secured to brick walls with galvanized mild steel pipe clip with barrel and back plate complete with nuts and bolts, including cutting and pinning.

Cost per 2·00 m length	233p
Take delivery	2p
One pipe clip	56p
Neoprene 0 ring joint	18p
	309p
Waste 5%	15p
Labour	
Plumbing pipe, making joint and fixing pipe clip	
50 minutes plumber @ 230p	192p
	516p
Profit and Oncost 20%	103p
2·00 m length costs	619p
Rate per m = 619 ÷ 2	£3·10

Example 12

Extra over 68 mm diameter rainwater pipe for offset bend with 306 mm projection.

Cost of offset	208p
Take delivery	2p
Neoprene O ring joint	18p
	228p
Waste 5%	11p
	239p
Labour	
Making joint and fixing (including cutting pipe if required)	
25 minutes plumber @ 230p	96p
C/fd	335p

	B/fd	335p
Profit and Oncost 20%		67p
		402p
Deduct 0·60 m of pipe at 310p per m		186p
		216p
Rate		£2·16

PVC SOIL AND WASTE PIPES

Example 13

110 mm PVC heavy grade soil and waste pipe with socket one end of each length, jointed with neoprene O ring and secured to brick walls with PVC coated mild steel socket pipe clip complete with nut and bolt, including cutting and pinning.

Cost per 2·00 m length	490p
Take delivery	2p
One pipe clip	79p
Neoprene O ring joint	18p
	589p
Waste 5%	29p
Labour	
Plumbing pipe, making joint and fixing pipe clip 55 minutes @ 230p	211p
	829p
Profit and Oncost 20%	166p
2·00 m length costs	995p
Rate per m = 995 ÷ 2	£4·98

Example 14

Extra over do. for 110 mm access bend

Cost of bend		484p
Take delivery		2p
	C/fd	486p

B/fd	486p
Neoprene O ring joint	18p
	504p
Waste 5%	25p
Labour	
Making joint, cutting pipe as required, and fixing	
40 minutes plumber @ 230p	153p
	682p
Profit and Oncost 20%	136p
	818p
Deduct 0·25 m of pipe @ 498p per m	124p
	694p
Rate	£6·94

Example 15

Extra over do. for 110 mm access branch	
Cost of branch	561p
Take delivery	2p
Two neoprene O ring joints	36p
	599p
Waste 5%	30p
Labour	
Making two joints, cutting pipe as required and fixing	
1 hour plumber @ 230p	230p
	859p
Profit and Oncost 20%	172p
C/fd	1,031p

	B/fd	1,031p
Deduct 0·39 m of pipe @ 498p per m		194p
		837p
Rate		£8·37

Example 16

42 mm PVC waste pipe, including straight couplings, fixed to brick wall with PVC coated mild steel clips at not exceeding 750 mm centres.

Note: PVC pipe is supplied in 4·00 m lengths.

Consider one 4·00 m length of 42 mm pipe @ £3·86 per length	386p
Take delivery	5p
Pipe clips; 4,100 ÷ 0·750 = 5·5 @ 13p	72p
Screws; 10 @ 125p per gross	9p
Straight coupling (1 per length) ∴ 1 @ 49p	49p
	521p
Allow for waste 2½%	13p
	534p
Labour	
Fitting up tube; plumber fits 3·00 m per hour @ 230p = 4·00 m ÷ 3·00 x 230p	307p
Fitting up straight coupling Plumber 15 minutes @ 230p	58p
	899p
Profit and Oncost 20%	180p
Rate per length of 4·00 m	1,079p
Rate per m = 1,079p ÷ 4·00	£2·70

Example 17

Extra for 42 mm bent coupling

42 mm bent coupling costs		65p
Waste 2½%		2p
	C/fd	67p

B/fd	67p
Labour	
Fixing; plumber 12 minutes @ 230p (including taking delivery)	46p
	113p
Profit and Oncost	23p
Rate	£1·36

Note: No deduction taken for girth of fitting as it is regarded as negligible.

Example 18

Extra for 42 mm tee.

42 mm tee costs	103p
Waste 2½% say	2p
Labour	
Fixing; plumber 18 minutes @ 230p (as above)	69p
	174p
Profit and Oncost 20%	35p
Rate	£2·09

Example 19

42 mm polypropylene bottle trap with 'P' outlet and 76 mm seal.

Cost of trap	133p
Waste 2½% say	3p
Labour	
Fixing; plumber 15 minutes @ 230p (as above)	58p
	194p
Profit and Oncost 20%	39p
Rate	£2·33

COPPER WASTE PIPE

Example 20

42 mm copper waste pipe to BS 2871 table 2 including straight couplings fixed to brick wall with brass holderbats at not exceeding 1·20 m centres.

Consider one 6·00 m length of 42 mm tube @ £1·48 per m	888p
Take delivery	4p
Holderbats: 6·00 ÷ 1·2 = 5 @ 80p each	400p
Brass screws: 10 @ 360p per gross	25p
Straight couplings (1 per length) ∴ 1 @	260p
	1,577p
Allow for waste $2\frac{1}{2}\%$	39p
Labour	
Fitting up tube: plumber fits 3·00 per hour = 6·00 ÷ 3·00 x 230p	460p
Fitting straight coupling: plumber 15 minutes @ 230p	58p
	2,134p
Profit and Oncost 20%	427p
Rate per 6·00 m	2,561p
Rate per m	£4·27

Note: The time for fixing
(*a*) bent coupling is 15 minutes plumber;
(*b*) tee coupling is 22 minutes plumber.

HOT AND COLD WATER INSTALLATIONS

Copper tubing is now purchased by 100 linear metres.

The labour output for fixing 15 mm, 22 mm and 28 mm tubing is the same for each of these sizes (although some estimators consider that the output varies for each size).

In the larger sizes the output would reduce according to size but it is suggested that 35 mm and 42 mm would be priced on the same output basis. And 54 mm would be priced on a lower output basis.

This also applies to fittings such as bends, tees, crosses, etc. There is a slight variation in the labour for forming bends as will be shown.

In accordance with the S.M.M. sixth edition, straight couplings are included in the measurement of tubing. Copper tubing is supplied in 6-m lengths. In practice the pipe runs are divided into shorter lengths to accommodate intervening fittings. An allowance of 1 straight connector per $1\frac{1}{2}$ lengths of tubing is a fair average, i.e. per 9 m.

Example 21

28 mm solid drawn copper tube to BS 2871 Table X fixed to timber with tinned steel clasps at not exceeding 1·20 m centres including straight couplings.

Consider one 6·00 m length of 28 mm tube @ £1·00 per m	600p
Take delivery	5p
Steel clasps: 6·00 ÷ 1·2 = 5 @ 50p per ten.	25p
Screws: 10 @ 140p per gross	10p
Straight couplings: 1 per 9·00 m; 1 coupling costs 129p ∴ $\frac{2}{3}$ of 129 p per 6·00 m length	86p
	726p
Allow for waste $2\frac{1}{2}$%	18p
	744p
Labour	
Fitting up tube: plumber fits 3·5 m per hour @ 230p = 6·00 ÷ 3·5 x 230p	394p
Fitting straight coupling: plumber take 10 minutes per coupling ∴ $\frac{2}{3}$ (230p ÷ 6)	25p
	1,163p
Profit and Oncost 20%	233p
Rate per 6·00 m	1,396p
Rate per m	£2·33

Example 22

22 mm solid drawn copper tube to BS 2871 Table X fixed to timber with tinned steel clasps at not exceeding 1·20 m centres including straight couplings.

Consider 6·00 m length of 22 mm tube @ 79p per m	474p
Take delivery	5p
Steel clasps: 6·00 ÷ 1·2 = 5 @ 29p per ten	15p
Screws: 10 @ 140p per gross	10p
Straight couplings: 1 per 9·00 m; one coupling costs 87p ∴ $\frac{2}{3}$ of 87p for 6·00 m length	58p
	562p
Allow for waste $2\frac{1}{2}$%	14p
	576p
Labour a.b.	419p
	995p
Profit and Oncost 20%	199p
Rate per 6·00 m	1,194p
Rate per m	£1·99

Example 23

Form bends on 15 mm, 22 mm and 28 mm copper tube.

	15 mm	22 mm	28 mm
Labour			
Forming bends on 15 mm plumber $7\frac{1}{2}$ minutes @ 230p	29p		
Forming bends on 22 mm plumber 10 minutes @ 230p		38p	
Forming bends on 28 mm plumber 12 minutes @ 230p			46p
	29p	38p	46p
Profit and Oncost 20%	6p	8p	9p
Rates are	35p	46p	55p

Example 24

Extra over copper tube for
(*a*) 15 mm bent coupling (compression type)
(*b*) 22 mm bent coupling (compression type)
(*c*) 28 mm bent coupling (compression type)

	15 mm	22 mm	28 mm
Bent couplings	80p	116p	167p
Waste $2\frac{1}{2}$%	2p	3p	4p
	82p	119p	171p
Labour			
Fixing: plumber 10 minutes @ 230p	38p	38p	38p
	120p	157p	209p
Profit and Oncost 20%	24p	31p	42p
Rates are	144p	188p	251p

Note: No deduction taken for girth of fitting.

Example 25

Extra over copper tube for

(*a*) 15 mm tee (compression type)
(*b*) 22 mm tee (compression type)
(*c*) 28 mm tee (compression type)

	15 mm	22 mm	28 mm
Tee costs	106p	155p	241p
Waste $2\frac{1}{2}$%	3p	4p	6p
	109p	159p	247p
Labour			
Fixing: plumber 15 minutes @ 230p	58p	58p	58p
	167p	217p	305p
Profit and Oncost 20%	33p	43p	61p
Rates are	200p	260p	366p

Note: (1) The fixing time for copper to iron adaptors is the same as for plain couplings.

(2) Capillary fittings normally take slightly longer than compression fittings; the suggested time for plumber fixing is:

(*a*) for bent couplings (15 mm, 22 mm, 28 mm), 15 minutes

(*b*) for tees (15 mm, 22 mm, 28 mm), 20 minutes.

(3) No deduction is normally made from copper tube for lengths of fittings.

Austenitic Stainless Steel tubes to BS 3014 are now being used instead of copper tubing in some situations. The prices are generally lower than copper (which is subject to volatile market fluctuation). The labour factors for stainless steel tube are the same as for equivalent sizes in copper, and capillary or compression fittings are the same as for copper.

Example 26

28 mm stainless steel, to BS 3014, fixed to timber with tinned steel clasps at not exceeding 1·20 metres centres including straight couplings.

Consider one 6·00 m length of 28 mm tube @ 81p per m	486p
Take delivery	5p
Steel clasps; 6·00 ÷ 1·2 = 5 @ 50p per ten	25p
Screws: 10 @ 140p per gross	10p
Straight couplings; 1 per 9·00 m; one coupling costs 129p	
∴ ⅔ of 129p per 6·00 m length	86p
	612p
Allow for waste 2½%	15p
	627p
Labour	
Fitting up tube; plumber fits 3·5 m per hour @ 230p = 6·00 ÷ 3·5 x 230p	394p
Fitting straight coupling; plumber takes 10 minutes per coupling	
∴ ⅔ (230p ÷ 6)	25p
C/fd	1,046p

	B/fd	1,046p
Profit and Oncost 20%		209p
Rate per 6·00 m		1,255p
Rate per m		£2·09

Example 27

22 mm do do

Consider one 6·00 m length of 22 mm tube at 64p per metre	384p
Take delivery	5p
Steel clasps; 6·00 ÷ 1·2 = 5 @ 29p per ten	15p
Screws; 10 @ 140p per gross	10p
Straight couplings; 1 per 9·00 m; one coupling costs 87p	
∴ $\frac{2}{3}$ of 87p for 6·00 m length	58p
	472p
Allow for waste $2\frac{1}{2}$%	12p
	484p
Labour	
as before	419p
	903p
Profit and Oncost 20%	181p
Rate per 6·00 m	1,084p
Rate per m	£1·81

Note: Labours such as forming bends are the same as for copper tube.

SANITARY FITTINGS

Sanitary fittings may be included in the Bill of Quantities as individual items for supply and fix. The alternative is to provide a Prime Cost Sum for the supply of sanitary fittings and to have individual items for fixing.

Example 28

Low level W.C. suite complete	4,800p
Offload, assemble, and fix W.C. suite complete	
4 hours plumber @ 230p	920p
	5,720p
Profit and Oncost 20%	1,144p
	6,864p
Rate	£68·64

Example 29

Pedestal wash basin size 500 x 400 mm with waste fitting and pillar taps	2,150p
Offload, assemble, and fix wash basin complete:	
3 hours plumber @ 230p	690p
	2,840p
Profit and Oncost 20%	568p
Rate	£34·08

Example 30

Rectangular bath size 1,800 mm long x 800 mm wide with taps, waste, with fibre glass front and one end panel complete	9,850p
Offload, assemble, and fix bath complete:	
7 hours plumber @ 230p	1,610p
	11,460p
Profit and Oncost 20%	2,292p
	13,752p
Rate	£137·52

16 Plaster etc. Works

Where ½ labourer is taken in the following examples, it is understood that the areas are big enough to warrant the time of a labourer moving from one tradesman to another. The mixing is included in the labourer's time and the cost of machinery priced for in the Preliminaries Bill.

PLASTERBOARD

British Gypsum Ltd is the sole manufacturer, at present, of plasterboard in the United Kingdom. The thickness of plasterboard types will be 9·5, 12·7 and 19 mm. The metric sizes conform to BSR 30 Table 1 as amended in 1973 generally as recommended for co-ordinating spaces for rigid sheet materials given in PD 6444 Appendix E.

Type

(*a*) *Wallboard* is plasterboard encased in paper with one ivory surface for direct decoration and has three types of joint:

(i) tapered for flush jointing;
(ii) square edge for cover strip jointing;
(iii) bevelled edge for V-jointing.

Insulating (foil-backed) wallboard is normally supplied with an ivory surface. It is available in 9·5 and 12·7 mm thicknesses in 600, 900, and 1,200 mm widths and 1,800 to 3,000 mm lengths for 12·7 mm thickness is recommended. Noggins will have to be used if fixed at maximum centres.

(*b*) *Lath* is a narrow-width plasterboard with rounded long edges and grey paper lining to both sides suitable to receive plaster. Joints do *not* require scrimming. It is available in 9·5 and 12·7 mm thicknesses in 406 x 1,200 mm sheets. Timber supports to be at maximum of 450 mm centres for 9·5 mm thickness and 600 mm centres for 12·7 mm thickness.

(*c*) *Baseboard* is a square-edge plasterboard obtainable in 9·5 mm thickness and in 914 x 1,200 mm sheets used as a base for plaster with scrimmed joints. Baseboard should be supported at all ends and edges at the perimeter of the ceiling, with joists at centres not exceeding 450 mm.

(*d*) *Plank* is used mainly for partitions and in framed construction where a high standard of sound insulation and fire resistance is required. Obtainable in 19 mm thickness in 600 mm width and 2,350 to 3,000 mm lengths. There are two types, one similar to wallboard with tapered or square edges and the other like lath with square edges only. Framing should be spaced at centres not exceeding 900 mm for ceiling and 750 mm centres for partitions, etc.

Cost

The unit rate will be expressed in £ per 100 m². Prices can vary according to changes in discounts allowed on load sizes by suppliers. The rates used are to be taken as averages.

Type	Thickness	Approx. cost d/d site in £ per 100 m²
Wallboard	9·5 mm	£59·38
	9·5 mm insulating	£70·44
	12·7 mm	£70·23
	12·7 mm insulating	£81·30
Lath	9·5 mm	£53·91
	9·5 mm insulating	£65·00
	12·7 mm	£66·48
	12·7 mm insulating	£77·54
Baseboard	9·5 mm	£53·91
	9·5 mm insulating	£65·00
	19·0 mm	£106·25
	19·0 mm insulating	£117·30

Waste Allow 5% waste for all plasterboard except lath where 2½% should be sufficient.

Example 1

9·5 mm Gypsum wallboard on ceilings fixed to joists at 450 mm centres with 30 mm No. 12 gauge galvanized flat headed nails at 150 mm centres and scrimmed joints.

Basic Rates
Wallboard in 1,800 x 900 mm sheets £59·38 per 100 m^2
Nails 65p per kg (110 per kg)
Scrim 120p per 100 m
Board finish £26·00 per tonne

Materials

Wallboard: £59·38 per 100 m^2 ∴ 1 m^2	60p
Take delivery and stick 135 m^2 per hour @ 190p	1½p
Nails: 1,800 x 900 mm sheet = 1·62 m^2	
Length of sheet 1,800 mm. Nos. required	
$\frac{1,800}{150} + 1 = 13$	
Width of sheet 900 mm. Nos. required	
$\frac{900}{450} + 1 = 3$	
No. of nails per sheet = 13 x 3 = 39 (if struts at 900 mm centres add 12 to make 51)	
Weight of nails required is $\frac{39}{1 \cdot 62 \times 110}$	
(No. of nails per kg) = 0·22 kg @ 65p per kg	14½p
Scrim: Half perimeter of board 2,700 mm	
∴ Length per m^2 = $\frac{2,700}{1 \cdot 62}$ = 1·70 m @ 120p	
per 100 m	2p
	78p
Allow 5% waste	4p
	82p
Board Finish: Allow 25 kg per 30 m of joint (including waste)	
∴ 1·70 m of joint per m^2 @ £26·00 plus 1 hour per tonne to take delivery and stack @	
C/fd	82p

	B/fd	82p
190p per hour = $\frac{2{,}790 \times 25 \times 1{\cdot}70}{1{,}000 \times 30}$		4p

Labour

Squad of 2 tradesmen and 1 labourer will fix 10 m² per hour

2 tradesmen @ 220p per hour	440p	
1 labourer @ 190p per hour	190p	
	630p	
∴ 1 m²		63p
Filling joints: squad does 30 m² per hour @ 63p per hour, ∴ 1 m²		21p
		170p
Profit and Oncost 20%		34p
		204p
Rate per m²		£2·04

Example 2

9·5 mm Gypsum baseboard on walls fixed to timber studs at 450 mm centres with 30 mm No. 12 gauge galvanized flat headed nails at 150 mm centres with scrimmed joints.

Basic Rates

Baseboard in 1,200 x 914 mm sheets	£53·91 per 100 m²
Nails	65p per kg
Scrim	120p per 100 m
Board finish plaster	£26·00 per tonne

Material

Baseboard: £53·91 per 100 m² ∴ 1 m²	54p
Take delivery and stack 135 m² per hour @ 190p ∴ 1 m²	1½p

Nails: 1,200 mm x 914 mm = 1·1 m²

Length of sheet is 1,200 mm: Nos. required =

$$\frac{1{,}200}{150} + 1 = 9$$

C/fd 55½p

	B/fd	$55\frac{1}{2}$p
3 rows of nails in width ∴ No. of nails per sheet = 3 x 9 = 27		
Weight of nails required is $\dfrac{27}{110 \times 1{\cdot}1}$		
= 0·22 kg @ 65p per kg		$14\frac{1}{2}$p
Scrim: Half perimeter of board is 2,114 mm		
∴ Length per m² = $\dfrac{2{,}114}{1{\cdot}1}$ = 1·92 m		
@ 120p per 100 m		2p
		72p
Allow 5% waste		4p
Board Finish: Allow 25 kg per 30 m of joint ∴ 1·92 of joint per m² @ £27·90 per tonne (see previous example)		
= $\dfrac{2{,}790\text{p} \times 25 \times 1{\cdot}92}{1{,}000 \times 30}$ = say		$4\frac{1}{2}$p
Labour		
Squad of 2 tradesmen and 1 labourer will fix 10 m² per hour		
∴ Cost of 1 m² as Example 1		63p
Filling joints and scrimming as Example 1		21p
		$164\frac{1}{2}$p
Profit and Oncost 20%		33p
		$197\frac{1}{2}$p
Rate per m²		£1·98

Example 3

9·5 mm Gypsum lath on walls fixed to timber studs at 400 mm centres with 30 mm No. 12 gauge galvanized flat headed nails at 150 mm centres with butt joints filled with neat plaster.

Basic Rates

Lath in 1,200 mm x 406 mm sheets	£53.91 per 100 m²
Nails	65p per kg
Board finish plaster	£26.00 per tonne

Material

Lath: £53·91 per 100 m² ∴ 1 m² — 54p

Take delivery and stack 135 m² per hour @ 190p ∴ 1 m² — $1\frac{1}{2}$p

Nails: 1,200 mm x 406 mm = 0·49 m²

Length of sheet is 1,200 mm: Nos. required

$$\frac{1{,}200}{400} + 1 = 4$$

Width of sheet is 406 mm: Nos. required

$$\frac{406}{150} + 1 = 4$$

∴ No. of nails per sheet = 4 x 4 = 16

Weight of nails required is

$$\frac{16}{0{\cdot}49 \times 110} = 0{\cdot}29 \text{ kg @ 65p per kg = say}$$ — 19p

— $74\frac{1}{2}$p

Allow $2\frac{1}{2}$% waste — 2p

Joints: (*Note:* No scrimming is required but the joints must be filled with neat plaster.)

Half perimeter of board is 1,606 mm

$$\therefore \text{Length per m}^2 = \frac{1{,}606}{0{\cdot}49} = 3{\cdot}275 \text{ m}$$

The amount of plaster required is 20 kg for anything from 30 to 35 m of joint

∴ 3·275 m of Board Finish with 20 kg per 35 m @ £27·90 per tonne

$$\text{(including stacking)} = \frac{20 \times 3{\cdot}275 \times £27{\cdot}90}{35 \times 1{,}000} = \text{say}$$ — 5p

C/fd $81\frac{1}{2}$p

	B/fd $81\frac{1}{2}$p
Labour	
Squad of 2 Tradesmen and 1 labourer will fix 10 m² per hour	
∴ Cost of 1 m² as Example 1	63p
Filling joints 60 m² per hour @ 630p	$10\frac{1}{2}$p
	155p
Profit and Oncost 20%	31p
	186p
Rate per m²	£1·86

Example 4

Two coats hardwall plaster 9·6 mm thick on gypsum baseboard walls.

Basic Rates

Washed sand	210p per tonne
Hydrated lime	£28.00 per tonne
Haird plaster	£26·55 per tonne
Finish plaster	£26·25 per tonne

Materials

1st coat: 8 mm thick, 1 haired plaster: $1\frac{1}{2}$ sand by volume (1 : $2\frac{1}{4}$ by weight) covers 205 m² per 1,000 kg

2nd coat: 1·6 mm thick, 1 finish plaster: 25% putty lime covers 370 to 410 m² per 1,000 kg

1st coat	
1 tonne of haired plaster @ £26·55	2,655p
Take delivery and stack $1\frac{1}{4}$ tonnes per hour @ 190p	152p
	2,807p
Mix 1 tonne haired plaster @ £28·07 per tonne C/fd	2,807p

B/fd	2,807p	
$2\frac{1}{4}$ tonnes sand @ 210p per tonne	473p	
	3,280p	
This mix (of 1 : $2\frac{1}{4}$ mix) covers 205 m² ∴ 1 m² = 3,280p ÷ 205		16p

2nd or finishing coat

Convert hydrated lime to putty lime	2,800p	
Take delivery and stack $1\frac{1}{4}$ tonnes per hour @ 190p	152p	
Convert to putty lime 2 hours @ 190p	380p	
	3,332p	
1 tonne finish plaster d/d site	2,625p	
Take delivery and stack $1\frac{1}{4}$ tonnes per hour @ 190p	152p	
	2,777p	
0·67 m³ @ 0·89 m³ per tonne of finish plaster @ 2,777p	2,090p	
0·23 m³ @ 1·67 m³ per tonne of lime @ 3,332p	459p	
	2,549p	

(0·89 m³ of finish plaster covers 370 m²)
(0·23 m³ of lime is 25% of 0·89 m³)
This mix covers approximately 370 m² 1·6 mm thick

∴ Cost per m² = $\frac{2,549p}{370}$ 7p

Labour

Squad of 2 tradesmen and 1 labourer will produce $5\frac{1}{2}$ m² per hour

∴ Cost per m² = $\frac{630p}{5\frac{1}{2}}$ 115p

C/fd 138p

	B/fd	138p
Profit and Oncost 20%		28p
		166p
Rate per m²		£1·66

Note: 2 coats hardwall plaster 9·6 mm thick on gypsum lath would be same rate per m²

Example 5

Two coats hardwall plaster 13 mm thick to brick walls

Basic Rates
Browning £26·55 per tonne
Finish plaster £26·25 per tonne
Washed sand 210p per tonne
Hydrated lime £28·00 per tonne

Mix

1st Coat: 11·1 mm thick 1 browning to 3 sand by volume (1 : 4½ by weight) covers 225 to 250 m² per 1,000 kg of browning.

Finishing Coat: 1·6 mm thick 1 finish to 25% putty lime covers 370 to 410 m² per 1,000 kg of finish.

Materials

1st coat: 1 tonne browning @ £26·55	2,655p
Take delivery and stack 1¼ tonnes per hour @ 190p	152p
1 tonne browning	2,807p
4½ tonnes washed sand @ 210p	945p
	3,752p

This mix covers approximately 225 m² 11·1 mm thick

$\therefore$ Cost per m² $= \dfrac{3{,}752\text{p}}{225}$ say 17p

C/fd 17p

B/fd 17p

Finishing coat:
See Example 4

0·67 m^3 @ 0·89 m^3 per tonne of finish @ 2,777p 2,090p
0·22 m^3 @ 1·67 m^3 per tonne of lime @ 3,332p 459p

2,549p

This mix covers approximately 370 m^2 1·6 mm thick

$\therefore$ Cost per $m^2 = \frac{2{,}549p}{370}$ 7p

Labour
Squad of 2 tradesmen and 1 labourer will produce 4½ m^2 per hour
$\therefore$ Cost per m^2 = 2 tradesmen @ 220p 440p
1 labourer @ 190p 190p

630p

$= \frac{630p}{4\frac{1}{2}}$ 140p

164p
Profit and Oncost 20% 33p

197p

Rate per m^2 £1·97

Example 6

Three coats hardwall plaster 19 mm thick to brick walls.

Mix

1st coat: 11·1 mm thick 1 browning: 3 sand by volume (1 : 4½ by weight)

2nd coat: 6·3 mm thick 1 browning: 3 sand by volume (1 : 4½ by weight)

Finishing coat: 1·6 mm thick 1 finish: 25% putty lime.

Basic Rates: As Example 5

Materials

1st coat: As Example 5	17p
2nd coat: Cost per m² *pro rata* is $\frac{6{\cdot}3}{11{\cdot}1} \times 17\text{p}$, say	10p
Finishing coat: As Example 5	7p

Labour
Squad of 2 tradesmen and 1 labourer will produce $3\frac{1}{2}$ m² per hour

∴ Cost per m² = $\frac{630\text{p}}{3\frac{1}{2}}$ (*see* Example 5)	180p
	214p
Profit and Oncost 20%	43p
	257p
Rate per m² is, say	£2·57

Example 7

Form external angle on plaster

Tradesman forms 16·50 m per hour @ 220p	$13\frac{1}{2}$p
Profit and Oncost 20%	$2\frac{1}{2}$p
Rate per m	16p

Note: The labour content for pencil rounded arises or internal angles would be the same as above.

Example 8

Two coats Carlite plaster 13 mm thick to brick walls.

Basic Rates
Carlite browning £35·50 per tonne
Carlite finish £27·55 per tonne

Material

Carlite browning per tonne	3,550p
Unload and stack $1\frac{1}{4}$ tonnes per hour @ 190p	152p
	3,702p

Carlite finish per tonne	2,755p
Unload and stack $1\frac{1}{4}$ tonnes per hour @ 190p	152p
	2,907p

1st coat: 11·1 mm thick
1 tonne Carlite browning covers 140 m²

$\therefore 1\ \text{m}^2 = \frac{3{,}702\text{p}}{140}$ $26\frac{1}{2}$p

Finishing coat: 1·6 mm thick
1 tonne Carlite finish covers 410 to 490 mm²

$\therefore 1\ \text{m}^2 = \frac{2{,}907\text{p}}{410}$ 7p

Labour
Squad of 2 tradesmen and 1 labourer will produce $4\frac{1}{2}$ m² per hour

$\therefore \text{Cost per m}^2 = \frac{630\text{p}}{4\frac{1}{2}}$

	140p
	$173\frac{1}{2}$
Profit and Oncost 20%	$34\frac{1}{2}$p
	208p
Rate per m²	£2·08

Example 9

Two coats cement plaster 19 mm thick on brick walls (1 : 3 mix).

Basic Rates		
Cement £24·00 per tonne		
Washed sand 210p per tonne		
Cement d/d site per tonne	2,400p	
Unload and stack $1\frac{1}{4}$ tonnes per hour @ 190p	152p	
	2,552p	
Materials per cubic metre		
Cement 1·44 tonnes @ 2,552p per tonne	3,675p	
Washed sand 1·52 tonnes @ 210p per tonne	319p	
Mix by volume		
1 m³ cement @ 3,675p	3,675p	
3 m³ sand @ 319p	957p	
	4,632p	
Allow 20% for reduction in bulk		
∴ add 25%	1,158p	
4 m³ costs	5,790p	
∴ 1 m³	1,448p	
∴ Cost per m² 19 mm thick is		
1,448p x 0·019		28p
Allow 5% waste		1p
Labour		
Squad of 2 tradesmen and 1 labourer will produce $5\frac{1}{2}$ m² per hour		
∴ 1 m² costs $\frac{630p}{5\frac{1}{2}}$		115p
	C/fd	144p

	B/fd	144p
Profit and Oncost 20%, say		29p
		173p
Rate per m^2		£1·73

Output on walls to receive tiles etc. for a squad of 2 tradesmen and 1 labourer:

One coat 13 mm thick: $7\frac{1}{2}$ m^2 per hour.
One coat 19 to 25 mm thick: $6\frac{1}{2}$ m^2 per hour.

Example 10

150 x 100 mm stucco gauged cornice having two lists at top and one at bottom, total girth 300 mm.

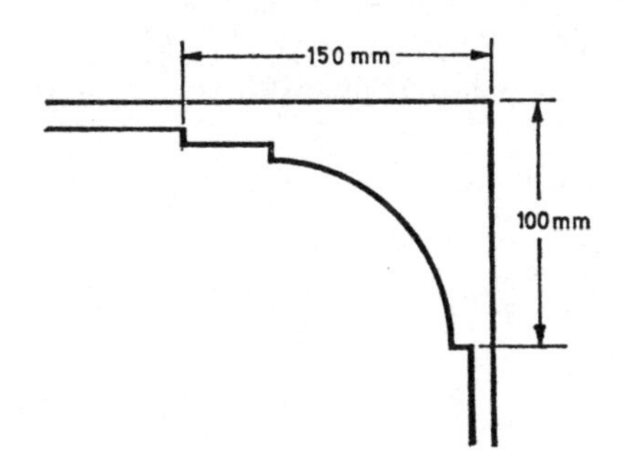

Material

Lime putty as before in Example 4		
= 3,332p per tonne		
0·60 tonne by volume is 1 m^3		
∴ 1 m^3 costs 3,332p x 0·60		1,999
Stucco per tonne costs £24·80	2,480p	
Take delivery and stack $1\frac{1}{4}$ tonnes per hour @ £1·90	152p	
	2,632p	

1 tonne by volume is 0·90 m^3

$\therefore 1\ m^3 = \frac{2,632}{0\cdot 90} \times 1$ 2,924p

C/fd 4,923p

B/fd	4,923p
Allow 10% for shrinkage	
∴ Add $\frac{1}{9}$	547p
Cost of 2 m³	5,470p

∴ 1 m³ costs 2,735p
Material required taking 0·03 m³ per m²
of cornice girth
1 m of 300 mm girth = 0·3 m²
∴ Material is 0·03 x 0·3 = 0·009 m³ per m
@ 2,735p per m³ say — 25p

Labour

Tradesman aligns ceiling and wall and fixes templates @ $3\frac{1}{2}$ m per hour	
∴ 1 m costs 220p ÷ $3\frac{1}{2}$	63p
2 tradesmen and 1 labourer produce 2 m per hour	
∴ 1 m costs 630p (*see* previous examples) ÷ 2	315p
	403p
Profit and Oncost 20%	81p
	484p
Rate per m	£4·84

SCREEDS

Cement Screeded Beds

Output by 2 tradesmen and 1 labourer:

Thickness	m² per hour
13 to 19 mm	$8\frac{3}{4}$
25 to 32 mm	$7\frac{1}{2}$
38 to 52 mm	$6\frac{1}{4}$

Cement Trowelled Beds

For linoleum, PVC, tiles, etc.; output by 2 tradesmen and 1 labourer:

Thickness	m² per hour
13 to 19 mm	5
25 to 32 mm	4½
38 to 52 mm	3¾

Example 11

25 mm cement floated bed (1 : 3) to receive PVC floor tiles.

Material

Cost of mix per m³ from Example 9 = 1,448p	
∴ Cost per m² 25 mm thick is 1,448p x 0·025	36p
Allow 5% waste, say	2p
Slurry	5p

Labour

Squad of 2 tradesmen and 1 labourer produce 4½ m² per hour	
∴ 1 m² costs 630p ÷ 4½	140p
	183p
Profit and Oncost 20%	37p
	220p
Rate per m²	£2·20

Granolithic

Output by 2 tradesmen and 1 labourer:

Thickness	m² per hour
up to 40 mm	3¾
40 to 60 mm	3
60 to 75 mm	2½

Example 12

25 mm Granolithic on concrete floor laid to falls.

This item will include preparing concrete with a coat of slurry (1 : 1).

Material

Mix: 1 cement : 2½ granite chips.

Granite Chips

1 tonne d/d site £6·30

1·44 tonnes = 1 m³

∴ Cost per m³ = 1·44 x 630p 907p

Cement

Cost from Example 9 per m^3	3,675p	
1 m^3 cement @ 3,675 p	3,675p	
$2\frac{1}{2}$ m^3 granite @ 907p	2,268p	
	5,943p	
Allow 20% reduction in bulk, ∴ add 25%	1,486p	
$3\frac{1}{2}$ m^3 costs	7,429p	
∴ 1 m^3	2,123p	
Material required per m^2, 25 mm thick is 2,123p x 0·025		53p
Allow 5% waste		3p
Slurry		5p

Labour

Squad of 2 tradesmen and 1 labourer will produce $3\frac{3}{4}$ m^2 per hour ∴ 1 m^2 costs 630p ÷ $3\frac{3}{4}$	168p
	229p
Profit and Oncost 20%	46p
	275p
Rate per m^2	£2·75

Example 13

19 mm granolithic projecting skirting 150 mm high with 15 mm radius cove next floor (backing measured elsewhere).

Material

Cost of mix per m^3, as previous example 2,123p	
∴ Cost of 1 m 19 mm thick and 150 mm high = 2,123p x 0·019 x 0·150	6p
Allow for material in cove and waste, say	2p
C/fd	8p

	B/fd 8p
Labour	
Forming skirting including temporary grounds at top and bottom:	
Squad of 2 tradesmen and 1 labourer will produce $3\frac{1}{2}$ m per hour	
$\therefore$ 1 m will cost 630p ÷ $3\frac{1}{2}$	180p
	188p
Profit and Oncost 20%	38p
	226p
Rate per m	£2·26

FLOOR QUARRY TILES

Floor quarry tiles are mainly sold in units of 1,000 in large quantities and in units of 10 in small quantities. The number of tiles per m² can be worked out as:

$$100 \times 100 \text{ mm tiles per m}^2 = \frac{1}{0{\cdot}100 \times 0{\cdot}100} = 100$$

$$150 \times 150 \text{ mm tiles per m}^2 = \frac{1}{0{\cdot}150 \times 0{\cdot}150} = 44{\cdot}44 \text{ say } 44\tfrac{1}{2}$$

$$200 \times 100 \text{ mm tiles per m}^2 = \frac{1}{0{\cdot}200 \times 0{\cdot}100} = 50$$

Outputs for 1 tilelayer and $\frac{1}{2}$ labourer:

Tile size	hours per m²
100 x 100 mm	$1\frac{1}{2}$
150 x 150 mm	1
150 x 75 mm	$1\frac{1}{3}$
Unload and stack	8 minutes per 100 for 1 labourer

Allow 0·015 m³ of cement and sand mix per m² of tiles. Allow 5% waste.

Example 14

100 x 100 x 22 mm quarry tiles to BS 1286 type A laid on screeded floor.

Materials

Cost of tiles per m² d/d site	335p
Unload and stack: 8 minutes @ 190p per hour	25p
	360p
Allow 5% waste	18p
0·015 m³ of cement and sand @ 1,448 per m³	22p

Labour

1 tradesman and ½ labourer will produce 1 m² in 1½ hours

1 tradesman	220p	
½ labourer	95p	
	315p x 1½	473p
		873p
	Profit and Oncost 20%	175p
		1,048p
	Rate per m²	£10·48

Example 15

150 x 150 x 22 mm quarry tiles to BS 1286 type A laid on screeded floor.

Materials

Cost of tiles per m² d/d site	370p
Unload and stack 4 minutes @ 190p per hour	12½p
	382½p
Allow 5% waste	19p
0·015 m³ of cement and sand @ 1,448p per m³	22p

Labour

1 tradesman and ½ labourer will produce 1 m² in 1 hour	315p
C/fd	738½p

	B/fd	$738\frac{1}{2}$p
Profit and Oncost 20%		$147\frac{1}{2}$p
		886p
Rate per m^2		£8·86

The analysis given above is for floors of a reasonable size not less than 6 m^2. For smaller areas, it is necessary to include for straight cutting by a tradesman at a rate of 30 minutes per m for all sizes of tiles.

Example 16

Billed quantity is 4 m^2; ∴ length of side = $\sqrt{4}$ = 2 m

Cutting need only be along two sides;
∴ Total length of cutting = 4 m
∴ Time to add for cutting = 4 x 30 minutes
= 2 hours

4 m^2 @ $738\frac{1}{2}$p (Example 15)	2,954p
Add for cutting 2 hours @ 220p	440p
	3,394p
Profit and Oncost 20%	679p
4 m^2 costs	4,073p
Rate per m^2	£10·18

Example 17 (Alternative to Example 15 illustrating quantities bought in large quantities by number).

150 x 150 x 22 mm quarry tiles bedded in cement mortar (1 : 3) and pointed with cement mortar (1 : 1).

Basic Rates
5,000 quarry tiles required @ £74·00 per 1,000
Mortar 1 : 3 @ 1,448p per m^3
Mortar 1 : 1 @ 2,497p per m^3
Transport £6·00 per tonne; 1,000 weigh
1·10 tonnes

Materials

5,000 tiles @ 7,400p per 1,000	37,000p
5·5 tonnes @ 600p per tonne	3,300p
Unload and stack 5 x $1\frac{1}{3}$ hours @ 190p per hour	1,267p
∴ 5,000 cost	41,567p
∴ $44\frac{1}{2}$ tiles per m² costs $\frac{41,567}{5,000} \times 44\frac{1}{2}$ = say	370p
i.e. 370p as against $382\frac{1}{2}$p	
Allow 5% waste	18p
Bed (1 : 3) 0·015 m³ per m² @ 1,448p per m³	22p
Pointing (1 : 1) 0·006 m³ per m² @ 2,497p per m³	15p
Labour	
1 tradesman and $\frac{1}{2}$ labourer as before (Example 15)	315p
Pointing 8 m² per hour @ 220p	28p
	768p
Profit and Oncost 20%	154p
	922p
Rate per m²	£9·22

CERAMIC WALL TILES

Co-ordinating sizes are 100 x 100 mm and 200 x 100 mm.

Thicknesses 4, 5, 6, and 10 mm.

Work sizes 98 x 98 mm and 197·5 x 98 mm, which allows for a width of joint of 1·5 mm and tolerances in BS 1281: 1966.

Output for 1 tilelayer and $\frac{1}{2}$ labourer:

Tile size	hours per m²
100 x 100 mm	2
150 x 150 mm	$1\frac{1}{2}$
200 x 100 mm	$1\frac{3}{4}$

Example 18

150 x 150 x 6 mm white wall tiles bedded in cement mortar (1 : 3) and pointed with Snowcrete.

Material	
Cost of tiles d/d site per m^2	315p
Unload and stack: 4 minutes @ 190p	13p
	328p
Allow 5% waste	16p
Bed (1 : 3) 0·015 m^3 per m^2 @ 1,448p per m^3	22p
Pointing with Snowcrete	21p
Labour	
1 tradesman and $\frac{1}{2}$ labourer will produce 1 m^2 in $1\frac{1}{2}$ hours @ 315p per hour (*see* Example 14)	473p
Pointing 8 m^2 per hour @ 220p	28p
	888p
Profit and Oncost 20%	178p
	1,066p
Rate per m^2	£10·66
Alternative	
Cost of tiles d/d site per 1,000	7,320p
Unload and stack: $1\frac{1}{2}$ hours @ 190p	135p
	7,455p
Allow 5% waste	373p
1,000 tiles cost	7,828p
$\therefore$ 1 m^2 @ $44\frac{1}{2}$ tiles = $\frac{7,828}{1,000} \times 44\frac{1}{2}$	348p
Bed as before	22p
Pointing as before	21p
Labour as before	501p
C/fd	892p

	B/fd	892p
Profit and Oncost 20%		178p
		1,070p
Rate per m²		£10·70

ROUGHCAST

Basic Rates

Cement	£24·00 per tonne
Sand	210p per tonne
Hydrated lime	£28·00 per tonne
Dorset peas	£11·25 per tonne
Wormit (Scotland) peas	£4·30 per tonne
Granite chips	£6·30 per tonne

The above rates would depend upon which area of the country the order was made in and could vary considerably. The dashes have been selected from a wide variety of chippings that are available from this and other countries.

Output for 2 tradesmen and 1 labourer:

Type	m² per hour
Dry dash	$4\frac{1}{3}$
Wet dash	$3\frac{3}{4}$

These figures are for plain work on fairly large areas and the output would drop by up to $\frac{1}{3}$ for small areas.

If applying wet roughcast, allow 2p per hour for tradesmen on areas over 40 m².

Materials

Mixes for examples: *1st coat:* 1 : 3; *2nd coat:* 1 : 1 : 6 (could be 1 : 3).

1 m³ cement costs 3,675p (Example 9).

1 m³ sand costs 319p (Example 9).

1 m³ lime costs 1,995p (Example 4 1·67 m³ per tonne)

Dorset peas: 1 m³ weighs 1·66 tonnes @ 1,125p per tonne

∴ 1·66 tonnes cost 1,125p × 1·66 = 1,868p

Wormit peas: 1 m³ weighs 1·92 tonnes @ 430p per tonne

∴ 1·92 tonnes cost 430p × 1·92 = 826p

Granite chips: 1 m³ weighs 1·60 tonnes @ 630p per tonne

∴ 1·60 tonnes cost 630p × 1·60 = 1,008p

Example 19

19 mm roughcast in three coats having Dorset pea finishing coat in dry dash.

Materials

1st coat

1 m³ cement @ 3,675p	3,675p	
3 m³ sand @ 319p	957p	
	4,632p	
Allow 20% reduction in bulk, ∴ add 25%	1,158p	
4 m³ costs	5,790p	
∴ 1 m³	1,448p	
1 m² 10 mm thick (allowing for waste) = 1,448 x 0·010		14½p

2nd coat

1 m³ cement @ 3,675p	3,675p	
1 m³ lime @ 1,995p	1,995p	
6 m³ sand @ 319p	1,914p	
	7,584p	
Allow 25% shrinkage for lime mix, ∴ add ⅓	2,528p	
8 m³	10,112p	
1 m³	1,264p	
1 m² 10 mm thick (allowing for waste) = 1,264p x 0·010		13p

Dorset peas

1 m³ covers 110 m² (from records) @ 1,868p per m³;

∴ 1 m² costs $\frac{1,868}{110}$ = say 17p

C/fd 44½p

	B/fd	$44\frac{1}{2}$p

Labour

2 tradesmen and 1 labourer will produce $4\frac{1}{3}$ m² per hour;

$\therefore$ 1 m² costs $\dfrac{630\text{p}}{4\frac{1}{3}}$ — $145\frac{1}{2}$p

Mixer

On a large contract the mixer cost would be priced for in the item rather than in the Preliminaries and would be shared between 3 or 4 squads.

Say cost of 452p per hour;

$\therefore$ Cost per m² $= \dfrac{452}{4\frac{1}{3} \times 4}$ = say — 26p

	216p
Profit and Oncost 20%	43p
	259p
Rate per m²	£2·59

Example 20

19 mm roughcast in three coats having granite chip wet dash finishing coat (1 : 2).

Material

1st coat as Example 19		$14\frac{1}{2}$p
2nd coat as Example 19		13p
Dashing coat		
1 m³ cement @ 3,675p	3,675p	
2 m³ granite chips @ 1,008p	2,016p	
	5,691p	
Allow 20% reduction in bulk		
$\therefore$ add $\frac{1}{4}$	1,423p	
3 m³	7,114p	
$\therefore$ 1 m³	2,371p	

1 m² say 6 mm thick costs 2,371p x 0·006 = say — 14p

C/fd	$41\frac{1}{2}$p

	B/fd	$41\frac{1}{2}$p
Labour		
2 tradesmen and 1 labourer will produce $3\frac{3}{4}$ m² per hour; ∴ 1 m² costs $\frac{634}{3\frac{3}{4}}$ (including 2p per tradesman)		169p
Mixer		
(as Example 19) $\frac{452p}{3\frac{3}{4} \times 4}$		30p
		$240\frac{1}{2}$p
Profit and Oncost 20%		48p
		$288\frac{1}{2}$p
Rate per m²		£2·89

Example 21

19 mm roughcast in three coats finished off with Wormit pea wet dash (1 : 2).

Material		
1st coat as Example 19		$14\frac{1}{2}$p
2nd coat as Example 19		13p
Dashing coat		
1 m³ cement @ 3,675p	3,675p	
2 m³ peas @ 826p	1,652p	
	5,327p	
Allow 20% reduction in bulk; ∴ add $\frac{1}{4}$	1,332p	
3 m³	6,659p	
∴ 1 m³	2,220p	
1 m² say 6 mm thick costs 2,220 x 0·006		$13\frac{1}{2}$p
	C/fd	41p

	B/fd 41p
Labour	
2 tradesmen and 1 labourer will produce	
$3\frac{3}{4}$ m² per hour; ∴ 1 m² costs $\frac{634}{3\frac{3}{4}}$	169p
	210p
Mixer	
As Example 20	30p
	240p
Profit and Oncost 20%	48p
	248p
Rate per m²	£2·48

Example 22

Two coats cement plaster 19 mm thick on brick walls to jambs of openings (1 : 3 mix).

Note: (1) The Sixth Edition of S.M.M. makes no provision for narrow widths but in the case of cement rendering to jambs of openings where the main areas are of roughcast we believe some distinction should be made.

(2) The labour factor should be increased by 25%.

Materials	
Cost per m² from Example 9 (including waste)	29p
Labour	
Squad of 2 tradesmen and 1 labourer will produce $5\frac{1}{2}$ m² per hour	
∴ 1 m² costs $\frac{630\text{p}}{5\frac{1}{2}}$	115p
Plus 25% additional labour	29p
	173p
Profit and Oncost 20%	35p
	208p
Rate per m²	£2·08

Example 23

Form external angles between roughcast and cement plaster jambs of openings.

Tradesman forms 16·5 m per hour @ 220p	13½p
Profit and Oncost 20%	2½p
	16p

17 Glazier Work

Glass for normal glazing can be obtained in large sheets or 'cut to size'. On most jobs it is cheaper to buy it cut. For other work it must be cut on site or in the glazier's shop.

Waste

5% waste on ready-cut glass; 10% waste on uncut glass.

Putty

Composed of whiting and raw linseed oil, mixed in the proportion of 50 kg whiting to 9 litres of linseed oil and left for 12 hours. Slight variations of this mix and the addition of other ingredients give greater durability or workability as desired.

A rule of thumb used in the trade is 0·09 kg of putty for every metre girth of glass and this is fairly accurate but for one or two exceptional cases when it is in excess of the amount required.

Average amount of putty required for both timber and steel casements:

Panes n.e. 0·10 m^2	1·25 kg per m^2
Panes exceeding 0·10 m^2 but n.e. 0·50 m^2	0·625 kg per m^2
Panes exceeding 0·50 m^2 n.e. 1·00 m^2	0·5 kg per m^2
Panes exceeding 1·00 m^2	0·46 kg per m^2

Putty required for metal sashes is a special kind and more expensive than linseed oil putty.

CLEAR SHEET GLASS

Sheet glass is the most economical transparent glass for general glazing purposes, but is now being superseded by float glass in certain thicknesses.

Qualities

OQ – Ordinary Glazing Quality. Suitable for general glazing purposes.

SQ – Selected Glazing Quality. For glazing work above ordinary quality.
SSQ – Special Selected Quality. For high grade work.

Thickness m	Dimensions Normal maximum sizes mm	
2	1,700 x 1,100	
3	2,100 x 1,300	
5	4·65 m^2	One dimension must not exceed 2,700 mm
5·5	9·3 m^2	
6	9·3 m^2	

The 2 mm glass is not normally recommended for glazing.

PATTERNED GLASS

Patterned glass has one flat surface and one textured, patterned or fluted giving varying degree of obscuration and diffusion. The patterns are divided into three groups in conformance with the sales price list called Group 1, Group 2, and Group 3. It is in two thicknesses, 3 and 5 mm, and the normal maximum sizes are 2,130 x 1,270 mm and 2,130 x 1,320 mm respectively.

ROUGHCAST GLASS

Thickness mm	Normal maximum sizes mm
5	3,710 x 1,270
6	3,710 x 1,270
10	3,710 x 1,270
12	4,420 x 2,490

OUTPUT FIGURES

(For sheet glass and most types of obscure glass)

	Hours of tradesman per m^2 for cut glass	
	Steel	*Timber*
Panes n.e. 0·10 m^2	2·15	1·50
Panes 0·10–0·20 m^2	1·75	1·33

	Hours of tradesman per m² for cut glass	
	Steel	*Timber*
Panes 0·20–0·50 m²	1·25	1·00
Panes 0·50–1·00 m²	1·00	0·75
Panes exceeding 1·00 m²	0·75	0·66

If the glazier has to cut the glass as well as glaze add 50%. Although the above table does not comply strictly with the S.M.M., the estimator would assess the average size of the panes and allow outputs to suit.

Prices for cut glass are more expensive than for glass not cut to size. Times shown include for unloading.

Example 1

3 mm SQ clear sheet glass in panes exc. 0·10 but n.e. 0·50 m² in wood casements (ready-cut glass).

Material	
Basic price of glass per m² d/d site	472p
0·625 kg putty @ £10·20 per 50 kg	13p
Glazing sprigs	5p
	490p
Allow 5% waste	25p
	515p
Labour	
1·00 hour of tradesman @ 220p	220p
	735p
Profit and Oncost 20%	147p
	882p
Rate per m²	£8·82

Example 2

3 mm figure rolled glass Group 1 in panes exc. 0·10 but n.e. 0·50 m² in wood casements (glass not cut to size).

Material	
1 m² of glass d/d site	230p
0·625 kg putty @ £10·20 per 50 kg	13p
Glazing sprigs	5p
	248p
Allow waste 10%	25p
	273p
Labour	
1 hour of tradesman @ 220p	220p
Add 50% for cutting	110p
	603p
Profit and Oncost 20%	121p
	724p
Rate per m²	£7·24

Example 3

3 mm OQ clear sheet glass in 60 panes each not exceeding 0·10 m² in metal casements.

Material	
1 m² of glass d/d site per m²	400p
1·25 kg putty @ £10·90 per 50 kg	27p
Steel sash clips	10p
	437p
Allow waste 5%	22p
Labour	
2·15 hours of tradesman @ 220p	473p
	932p
Profit and Oncost 20%	186p
	1,118p
Rate per m²	£11·18

CLEAR FLOAT, PLATE GLASS

Plate glass is now largely superseded by float glass. This is sold cut to sizes in various thicknesses. The larger the area the greater the cost per m². Exact and special thicknesses are quoted only at a special price. 2 to 6 men may be required for glazing large heavy sheets.

It can be obtained in sheets from 3 to 38 mm thick. The normal maximum size of 6 mm sheet is 4,600 x 3,200 mm.

Output figures for up to 4 mm thick are same as for sheet glass. Output figures for 6 mm thick are half those for sheet glass.

If bedded in putty with beads, allow 0·10 kg per m² and 5p per m² for glazing sprigs.

Allow 5% for waste.

Float glass is classified (according to manufacturer's tariff) as follows:

(i) not exceeding 4·00 m²,
(ii) not exceeding 9·30 m²,
(iii) over 9·30 m².

Example 4

4 mm Float glass in panes exceeding m² in steel casements (glass not cut).

Material	
Basic price of glass d/d site per m²	271p
0·46 kg putty @ £10·90 per 50 kg	10p
Steel sash clips	10p
	291p
Allow 10% waste (glass not cut)	29p
Labour	
0·75 hours @ 220p	165p
Add 50% for cutting	83p
	568p
Profit and Oncost 20%	114p
	682p
Rate per m²	£6·82

Example 5

6 mm float glass in panes exceeding 1 m² in timber sashes with beads (glass cut to size).

Material	
Basic price of glass d/d site per m²	870p
0·10 kg putty @ £10·20 per 50 kg	2p
Sprigs	5p
	877p
Allow 5% waste	44p
Labour	
1·32 hours @ 220p	290p
	1,211p
Profit and Oncost 20%	242p
	1,453p
Rate per m²	£14·53

Labours

Curved Cutting. Ranges from 1½ to 2½ lin.m per hour depending upon the type and thickness of glass.

Hacking out old Glass. Taken as 2¼ hours per m² for a glazier for all kinds of glass except polished plate glass which is taken as 3½ to 4 hours per m². This must then be converted into a linear metre rate by reference to the size of panes to be hacked out.

DOUBLE GLAZING BY HERMETICALLY SEALED UNITS

Double or multiple factory-made hermetically sealed units are measured by enumerating, stating the size and type of unit.

Three of the well known types in general use are:

(1) Pilkington's 'Insulight Glastoglas' double glazing units having an overall thickness of 11 mm with 5 mm air space. They are normally made in stock sizes for standard size windows only. Specified sizes are made to order.

(2) Pilkington's 'Insulight' MK VI double glazing units,

which may be of sheet, float, armourplate, rough cast, glass, etc., with air space which may vary from 5 to 12 mm as required. Sizes may be obtained as required.

(3) Plyglass, which is similar in all respects to 'Insulight' MK VI.

Example 6

Insulight MK VI double glazing unit comprising 6 mm float glass with 12 mm air space, in size 1,830 x 1,830 mm, and glazing in wood with compound, beads, and screws, (supplied by joiner).

Material	
Unit size 1,830 x 1,830 costs	7,300p
Bedding compound (3·0 kg per m) = 22 kg @ 1,210p per 50 kg	532p
	7,832p
Allow 5% for waste	392p
Labour	
4 hours @ 220p	880p
	9,104p
Profit and Oncost 20%	1,821p
	10,925p
Rate per unit	£109·25

Example 7

Insulight MK VI double-glazing unit comprising 4 mm sheet glass with 6 mm air space, in size 610 x 1,220 mm, and glazing in metal with compound and putty.

Material	
Unit size 610 x 1,220 mm costs	2,900p
Bedding compound (1·5 kg per m) = 5·5 kg @ £12·10 per 50 kg	133p
Steel sash putty (to front of sash) 5·5 kg @ £10·90 per 50 kg	120p
C/fd	3,153p

	B/fd	3,153p
Allow 5% for waste		158p
Labour		
1·5 hours @ 220p		330p
		3,641p
Profit and Oncost 20%		728p
		4,369p
Rate per unit		£43·69

PATENT GLAZING SYSTEMS

There are a large number of patent glazing systems on the market at present offering a wide variety of choice with glazing bars produced for a single or double glazing as required. The work may be done by the specialists who manufacture, supply, and fix, or the glazing contractor may purchase the glazing bars and fix on site using his own labour, supplying the glass himself.

A typical example of the glazier fixing, using his own labour, with proprietary aluminium glazing bars, is as follows.

Note: The height to eaves of roof from ground level is 6·00 m and the bearings are timber.

Example 8

Patent glazing to roof consisting of Messrs X and Co. (or equally approved type) aluminium glazing bars 2,130 mm long placed at 600 mm centres to a span of 1,930 mm with 6 mm Georgian wired cast glass including alignment one way and providing all necessary glass stops, weather fillets, and fittings complete.

Material		
6 mm Georgian wired cast glass per m^2		693p
Aluminium glazing bars at 600 mm centres require $\left(\frac{1,000}{600} \times 1{\cdot}00\right)$ + 20% (*see* Chapter 12, under diagram (pp. 245–6) = say 2·00 m of bar per m^2 @ 775p per length of 2,130 mm bar (including stops, fillets etc.)		873p
	C/fd	1,566p

	B/fd	1,566p
Allow 5% for waste		78p
		1,644p
Labour		
Setting bars and glass: 1 hour @ 220p		220p
		1,864p
Profit and Oncost 20%		373p
		2,237p
Rate per m²		£22·37

18 Painter Work

Painter work is difficult to price consistently on account of different covering capacities, different qualities of brushes, different absorption of the various surfaces, different abilities of tradesmen.

An example of this would be how well the surface of the plaster had been polished by the tradesman, which can make a considerable difference to the 'drag' on the paint brush.

Pricing is normally split into the following sections:

(1) Preparation.
(2) Undercoating.
(3) Finishing.

PREPARATION

According to the Sixth Edition of the Standard Method of Measurement the type of preparatory work should be clearly stated and it will be taken for our purposes that 'preparation' has been defined in the 'Preambles', and is deemed to be included in the item.

General preparation of new surfaces is included in the paint application rates.

	Tradesman hours per 100 m^2
Knotting	$3\frac{1}{2}$
Stopping	$3\frac{1}{2}$
Size	10
Stain	10
Varnish	15
Rubbing down and removing loose	
limewhite, whitening, distemper	$11\frac{1}{2}$
Washing off non-washable distemper	15 to 25

Scrubbing off washable distemper and cleaning of loose particles	15 to 35
Washing and rubbing down paintwork	15 to 30
Burning off paint and sandpapering plain surfaces	80 to 125

OUTPUTS

Cement and Emulsion Paints

Tradesman hours per 100 m²

	Unprimed surface	*Primed surface*
Cement paint	11	9
Emulsion	11	9

Oil Paint

Tradesman hours per 100 m²

	Steel	*Plaster*	*Wood*	*Brick*	*Stone/ Conc.*
Priming	15	16	15	25	20
Undercoat	14	15	14	15	15
Finish	15	16	15	17	17

For work on isolated surfaces not exceeding 300 mm girth allow the following increase on labour:

(i) not exceeding 150 mm girth $33\frac{1}{3}$%
(ii) exceeding 150 mm but not exceeding 300 mm girth 20%

If steel is primed in the factory, allow 5 hours for touching up.

Covering Capacities per 100 m²

Cement Paint	on Brick	10 kg
	on Plaster	15 kg
Emulsion Paint	on Brick	$8\frac{1}{2}$ litres
	on Plaster	8 litres

Oil Paint		
Primer	on Plaster on Wood on Metal	10 litres
	on Brick	14 litres
	on Stone/Conc.	13 litres
Undercoats	on Plaster on Wood on Metal	$8\frac{1}{2}$ litres
	on Brick	10 litres
	on Stone/Conc.	9 litres
Finishing coats	All surfaces	5 to 8 litres
Varnish	on Wood	8 litres
Stain on Wood	Oil Stain	7 litres
	Spirit Stain	$3\frac{1}{2}$ litres
Knotting		$\frac{3}{4}$ litres
Stopping	Putty	2·5 kg
	Glass paper	28 sheets

BRUSHES

The outlay on brushes can be considerable and can be allowed for thus:

(1) Allow 5% as an overall percentage over labour and material.
(2) Life of brush is approximately 100 hours.

If used $\frac{2}{3}$ of painter's time the brush will last 150 hours. Cost of good quality brush say 315p, Cost per hour = 315/150 = 2p say.

The 2p should be multiplied by tradesman's time and added into the calculation.

PROFIT AND ONCOST

From experience it has been found that the cost of overheads required to administer and manage this trade is not as high as other sections of the building industry. We have found that an allowance of $12\frac{1}{2}$% for Profit and Oncost is sufficient and this has been allowed for in the following examples.

Example 1

Two coats emulsion on plastered ceilings.

Material		
Emulsion paint @ £6·00 per 5 litres		
Two coats @ 16 litres per 100 m² =		1,920p
Labour		
1st coat	11 hours per 100 m²	
2nd coat	9 hours per 100 m²	
	20 hours @ 220p	4,400p
Brushes		
20 hours @ 2p		40p
		6,360p
	Profit and Oncost 12½%	795p
	100 m² costs	7,155p
	Rate per m², say	72p

Example 2

One coat primer, two coats undercoat and one coat gloss oil paint on woodwork generally.

Material	
Stopping: 2·5 kg of putty @ £9·00 per 50 kg	45p
28 sheets of glass paper @ £11·00 per 500	62p
Knotting: 0·75 litre @ £4·00 per 5 litres	60p
Priming: 10 litres @ £4·50 per 5 litres	900p
Two coats undercoat: 2 x 8½ @ £6·50 per 5 litres	2,210p
One coat gloss oil: 6 litres @ £6·65 per 5 litres	798p
C/fd	4,075p

		B/fd	4,075p
Labour			
Stopping	3·5 hours		
Knotting	3·5 hours		
Priming	15·0 hours		
Two coats undercoat 2 x 14	28·0 hours		
One coat gloss oil	15·0 hours		
	65·0 hours @ 220p		14,300p
			18,375p
Brushes 65 hours 2p			130p
			18,505p
	Profit and Oncost 12½%		2,313p
	100 m² costs		20,818p
	Rate per m²		£2·08

Example 3

One coat primer, two coats undercoat and one coat gloss oil paint on isolated surfaces of woodwork exceeding 150 mm but not exceeding 300 mm girth.

Where painter work is measured by metres, an overall percentage of approximately 20% should be added to the price per m² to cover for the greater amount of 'cutting to line', i.e. the extra labour involved in the joining of one type of finish to another.

Material			
From Example 2			4,075p
Labour			
From Example 2	14,300p		
plus 20%	2,860p		17,160p
Brushes as before			130p
		C/fd	21,365p

	B/fd	21,365p
Profit and Oncost 12½%		2,671p
100 m² costs		24,036p
∴ 1 m, 200 mm wide, costs $\frac{24,036}{100} \times 0{\cdot}20$		48p
Rate per m		48p

Example 4

Wash and rub down existing oil paint on woodwork generally and paint with one coat undercoat and one coat gloss oil finish.

Material		
Undercoating 8½ litres @ £6·50 per 5 litres		1,105p
Finishing coat 5 litres @ £6·65 per 5 litres		665p
Labour		
Wash and rub down	15 hours	
Undercoat	14 hours	
Finishing coat	15 hours	
	44 hours @ 220p	9,680p
Brushes 44 @ 2p		88p
		11,538p
Profit and Oncost 12½%		1,442p
100 m² costs		12,980p
Rate per m²		£1·30

Example 5

One coat cement paint on stone walls.

Material		
20 kg of cement paint @ £10·80 per 40 kg		540p
4·5 litres of petrifying liquid @ 90p per 5 litres		81p
Labour		
11 hours @ 220p		2,420p
	C/fd	3,041p

	B/fd	3,041p
Brushes 11 @ 2p		22p
		3,063p
Profit and Oncost $12\frac{1}{2}\%$		383p
100 m^2 costs		3,446p
Rate per m^2		35p

Example 6

One coat stain and two coats varnish on woodwork generally.

Material		
7 litres of stain @ £3·55 per 5 litres		497p
1·4 kg of size @ £1·40 per 50 kg		4p
Two coats of varnish 2 x 8 = 16 litres @ £6·05 per 5 litres		1,936p
Labour		
Size	10 hours	
Stain	10 hours	
Two coats of varnish 2 x 15	30 hours	
	50 hours @ 220p	11,000
Brushes 50 x 2p		100
		13,537
Profit and Oncost $12\frac{1}{2}\%$		1,692
100 m^2 costs		15,229
Rate per m^2		£1·52

PAPERHANGING

Wallpaper is sold by the piece or roll. A standard size roll is 530 mm by 10·06 m = 5·35 m^2. There is also foreign size wallpaper which does not form part of these notes.

The supply and delivery to site of paper, fabric or plastic wall coverings is now the subject of a Prime Cost Sum.

Waste

It is normal practice to allow 10% for plain papers and 10 to 15% for patterned paper. Wallpapers are of various kinds and the labour involved in hanging them is variable.

Types of Wallpaper

(1) Ordinary wallpapers; pattern printed on paper as made.
(2) Rough textured wallpapers: fibre ingrained.
(3) Embossed papers: patterns raised.
(4) Lining papers: underlay for the better quality papers.
(5) Washable papers: for bathroom, etc.
(6) Satin papers.

Allow 1 litre of paste per roll.

In preparing old surfaces any holes or cracks are filled with stopping.

Labour

	Per roll
Clean down and stop walls	$\frac{1}{4}$ hour
Strip old paper	$\frac{2}{3}$ hour
Size	$\frac{1}{4}$ hour
Hang ordinary wallpapers	1 hour
Hang embossed wallpapers	$1\frac{1}{2}$ hours
Borders 40 mm wide	20 m per hour

Example 7

Hang patterned wallpaper to plastered walls including matching patterns, sizing walls and pasting; in 40 pieces.

Materials

Paste and size 8p

Labour: per piece or roll

Size $\frac{1}{4}$ hour
Hang 1 hour
$1\frac{1}{4}$ hours @ 220p 275p

C/fd 283p

	B/fd	283p
Profit and Oncost $12\frac{1}{2}\%$		35p
Cost of hanging one piece of $5{\cdot}35\ m^2$		318p
$\therefore$ Rate per m^2 (hanging only) $= 318p \div 5{\cdot}35$		59p

19 Drainage

Since the first edition of this book was published (1972), the traditional form of underground drainage by stoneware and fireclay pipes, with spigot and socket, and cement mortar and rope yarn joints has been rapidly superseded by the use of vitrified clay pipes and fittings with flexible joints, to BS 65 and 540.

The flexible joints facilitate greater speed and ease in laying and the pipes are manufactured in longer lengths than the traditional 2·00″ (0·60 m) for stoneware and 3·00″ (0·91 m) lengths for fireclay, and are produced with spigot and socket joints or spigot ends jointed with separate sleeves, e.g. HEPSEAL or HEPSLEVE.

However, although the use of the traditional stoneware pipe is declining three examples of this method will be shown below.

STONEWARE PIPES

MORTAR JOINTS

Size of pipe	Amount of rope yarn	Amount of mortar
100 mm nominal	0·02 kg	0·0005 m^3
150 mm nominal	0·03 kg	0·0009 m^3
225 mm nominal	0·05 kg	0·0017 m^3

Labour

Laying and jointing pipes (1 jointer plus 1 labourer):

100 mm nominal pipe	8 minutes per pipe
150 mm nominal pipe	12 minutes per pipe
225 mm nominal pipe	15 minutes per pipe

The above times are for pipes laid in trenches n.e. 2·00 m deep. If deeper, multiply the above labour rates by the following:

2·00 to 4·00 m deep x 1·75
4·00 to 6·00 m deep x 2·5
If laid on surface x 0·7
Allow 5% for waste.

The working rule states the definition of labourer's work (which will be common to all regions). Para. (*c*) states labourers engaged on drain work are to be paid labourer's rates. In fact, they are paid an additional amount, sometimes as much as a tradesman's rate.

Example 1

100 mm, nominal stoneware pipes laid and jointed in trenches with rope yarn and cement mortar (1 : 1) joints. (Note. Stoneware pipes will be taken as 1·00 m long.)

Material		
Pipe (best quality) 1·00 m long. Each	67p	
Take delivery and stack	5p	
	72p	
Allow 5% waste	4p	
	76p	76p
Joint		
Mortar: mixing included in labourer's output		
1 m³ cement @ 3,675p	3,675p	
1 m³ sand @ 319p	319p	
	3,994p	
Allow 20% shrinkage ∴ add ¼	999p	
2 m³ costs	4,993p	
∴ 1 m³ costs	2,297p	
∴ 0·0005 m³ @ 2,297p per m³	1p	
Rope yarn 0·02 kg @ 190p per kg	4p	
	5p	5p
	C/fd	81p

		B/fd	81p
Labour			
1 layer and 1 labourer take 8 minutes per pipe			
1 layer say	220p		
1 labourer	190p		
	410p per hour		
∴ Cost for 8 minutes is $\frac{410}{60} \times 8$			55p
			136p
Profit and Oncost 20%			27p
Cost of 1·00 m length of pipe			163p
Rate per m			£1·63

Example 2

100 mm, nominal, stoneware pipe in 4 branches each not exceeding 3 m long.

Material	
Pipe As Example 1	76p
Labour	
As Example 1	55p
	131p
Allow 10% for extra labour and material (for short runs in branches)	13p
Joint As Example 1	5p
	149p
Profit and Oncost 20%	30p
Cost of 1·00 m length of pipe	179p
Rate per m	£1·79

Example 3

Extra over 100 mm, nominal, stoneware pipe for bend.

Material	
Bend cost	60p
Take delivery and stack	5p
	65p
Allow 5% waste	3p
Joint As Example 1	5p
Labour	
1 layer and 1 labourer will take 12 minutes including cutting @ 410p per hour	82p
	155p
Profit and Oncost 20%	31p
	186p
Deduct Girth of bend 0·30 m @ 163p (*see* Example 1)	49p
	137p
Rate	£1·37

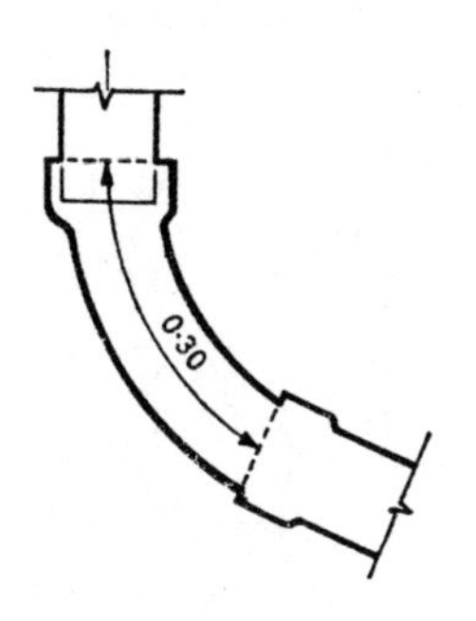

Note: This is the shortest girth of bend; other girths up to 610 mm are available.

Vitrified Clay Drainage Systems with Flexible Joints; to BS 65 and 540 Parts 1 and 2. Spigot and socket pipes are available in standard lengths as follows:

100 mm nominal diameter 1·25 m standard length
150 mm nominal diameter 1·50 m standard length
225 mm nominal diameter 2·00 m standard length
Larger diameters for sewer work are also available

The cost of pipes and fittings vary according to the locational Zone throughout the UK, there being 4 Zones, Zone 1 being the cheapest with the prices increasing slightly up to Zone 4.

Example 4

100 mm, nominal, vitrified clay pipes with approved spigot and socket flexible joints laid in trench.

Material	
Pipe in 1·25 m lengths, d/d Zone 1, cost per m is £1·03	
∴ cost per standard length of pipe is	129p
Take delivery and stack	5p
Joint lubricant; per joint	1p
	135p
Allow 5% waste	7p
Labour	
1 layer and 1 labourer take 4 minutes per pipe @ 410p	27p
	169p
Profit and Oncost 20%	34p
Cost per 1·25 m length	203p
Rate per m = $\frac{1000}{1250} \times 203$p	£1·62

Example 5

100 mm, nominal, vitrified clay pipes with approved spigot and socket flexible joints in 4 branches not exceeding 3 m long.

Material	
Pipe as Example 4	142p
C/fd	142p

	B/fd 142p
Labour	
As Example 4	27p
	169p
Allow 10% for extra material and labour (For short runs in branches)	17p
	186p
Profit and Oncost 20%	37p
Cost per 1·25 m length	223p

Rate per m = $\frac{1000}{1250}$ x 223p = £1·78

Example 6

Extra over 100 mm, nominal, vitrified clay pipe for bends.

Material	
Bend cost	175p
Take delivery and stack	5p
Joint lubricant	1p
	181p
Allow 5% waste	9p
Labour	
1 layer and 1 labourer will take 7 minutes including cutting @ 410p per hour	48p
	238p
Profit and Oncost 20%	48p
	C/fd 286p

B/fd	286p
Deduct	
Girth of bend 0·30 m @ 162p (*see* Example 4)	49p
	237p
Rate	£2·37

Example 7

Extra over 100 mm, nominal, vitrified clay pipe for branch.

Material	
Branch cost	222p
Take delivery and stack	5p
Joint lubricant	1p
	228p
Allow 5% waste	11p
Labour	
1 layer and 1 labourer will take 10 minutes (2 joints and 1 cut) @ 410p per hour	68p
	307p
Profit and Oncost 20%	61p
	367p
Deduct	
Length of branch plus arm 0·75 m @ 162p (*see* Example 4)	122p
	246p
Rate	£2·46

Example 8

150 mm, nominal, vitrified clay pipe with approved spigot and socket flexible joints laid in trench.

Material

Pipe in 1·50 m lengths d/d Zone 1, cost per m is £1·73

∴ cost per standard length of pipe is	260p
Take delivery and stack	6p
Joint lubricant; per joint	2p
	268p
Allow 5% waste	13p
Labour	
1 layer and 1 labourer take 6 minutes per pipe @ 410p	41p
	322p
Profit and Oncost 20%	64p
Cost per 1·50 m length	386p

Rate per m = $\frac{1000}{1500}$ x 386p = £2·57

Example 9

Extra over 150 mm, nominal, vitrified clay pipe for bend.

Material	
Bend cost	304p
Take delivery and stack	6p
Joint lubricant	2p
	312p
Allow 5% waste	16p
Labour	
1 layer and 1 labourer will take 10 minutes including cutting @ 410p per hour	68p
	396p
Profit and Oncost 20%	79p
C/fd	475p

B/fd	475p
Deduct	
Girth of bend 0·45 m @ 257p (*see* Example 8)	116p
	359p
Rate	£3·59

Example 10

Extra over 150 mm, nominal, vitrified clay pipe for branch.

Material	
Branch cost	370p
Take delivery and stack	6p
Joint lubricant	2p
	378p
Allow 5% waste	19p
Labour	
1 layer and 1 labourer will take 15 minutes including cutting at 410p	103p
	500p
Profit and Oncost 20%	100p
	600p
Deduct	
Girth of branch 0·75 m @ 257p (*see* Example 8)	193p
	407p
Rate	£4·07

Example 11

100 mm, nominal, vitrified clay LBS trap with 100 mm back inlet top piece and two 100 mm raising pieces to surface with stopper (cover seating is priced separately).

Material	
LBS trap	145p
Top piece	135p
Two raising pieces	104p
Stopper	70p
Joints lubricant, say	6p
Take delivery and stack	20p
	480p
Labour	
1 layer and 1 labourer will take 40 minutes @ 410p	274p
	754p
Profit and Oncost 20%	151p
	905p
Rate per m	£9·05

An alternative to spigot and socket pipes is the vitrified clay pipe with plain spigot ends which are coupled by polypropylene couplings. The pipes are available in 100 mm and 150 mm diameters only and the standard length is 1·60 m for both sizes. One example only is shown below.

Example 12

100 mm, nominal, vitrified clay pipes with plain spigot ends, jointed with approved polypropylene coupling, laid in trench.

Material	
Pipe in 1·60 m lengths, d/d Zone 1, cost per metre is £0·65	104p
Coupling	36p
Take delivery and stack	7p
Joint lubricant	1p
	148p
Allow 5% waste	7p
C/fd	155p

B/fd	155p
Labour	
1 layer and 1 labourer will take 8 minutes per pipe @ 410p	55p
	210p
Profit and Oncost 20%	42p
Cost per 1·60 m length	252p
Rate per m = $\frac{1000}{1600}$ x 252p =	£1·58

CAST IRON DRAINAGE

Example 13

100 mm, nominal, cast iron pipes to BS 437 laid and jointed in trenches with rope yarn and lead joints.

Material	
100 mm pipe; cost per 3·00 m length	1,460p
Take delivery and stack 5 minute labourer @ 190p	16p
Batt lead in joint 2·10 kg @ 25p per kg	53p
Rope yarn in joint 0·16 kg @ 36p per kg	6p
	1,535p
Allow waste 5%	77p
Labour	
Laying in trench and making joint 50 minutes plumber and labourer/mate per 3·00 m length @ 230p + 190p = 420p per hour	350p
	1,962p
Profit and Oncost 20%	392p
Cost per 3·00 m length	2,354p
Rate per m = 2354 ÷ 3	£7·85

Example 14

Extra over 100 mm, nominal, cast iron drain pipe for plain bend.

Material	
Bend cost	500p
Take delivery and stack 3 minute labourer @ 190p	10p
Batt lead in joint 2·10 kg @ 25p per kg	53p
Rope yarn in joint 0·16 kg @ 36p per kg	6p
	569p
Allow waste 5%	28p
Labour	
Laying in trench and making joint ½ hour plumber and labourer/mate @ 430p	215p
	812p
Profit and Oncost 20%	162p
	974p
Deduct Girth of bend 0·50 m @ 785p	392p
	582p
Rate	£5·82

Example 15

Concrete (1 : 3 : 6) 20 mm aggregate in surround to 100 mm internal diameter drain pipe, 440 mm total width and 440 mm total thickness.

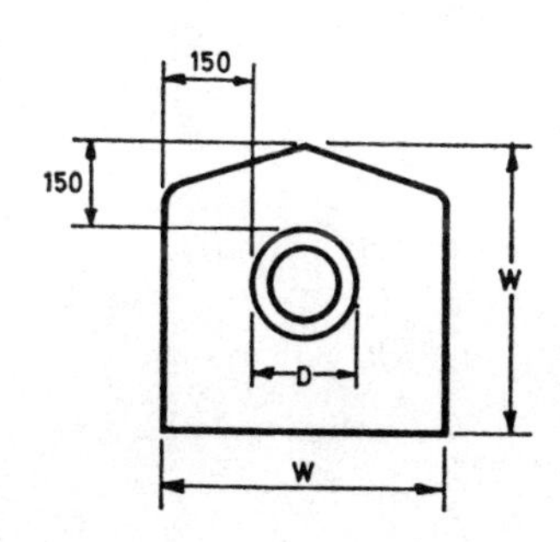

$W = D + 300$ mm where D is the external diameter of the pipe. Consider 1 m length:

Volume of concrete 0·44 x 0·44 x 1·00		0·194
(pipe wall 20 mm)		
Deduct pipe π/0·07 x 0·07 x 1·00	0·015	
shoulders 2/½/0·08 x 0·22 x 1·00	0·018	0·033
		0·161 m³

Cost of 1 m³ (1 : 3 : 6) concrete 20 mm aggregate

Materials for 1 m³ (*see* Chapter 6, Example 1)	977p
Mixing using 5/3½ mixer (*see* Chapter 7, Example 2)	452p
Transporting and placing 4½ hours @ 190p	855p
	2,284p
Profit and Oncost 20%	457p
Rate per m³	2,741p
∴ 1 m costs 2,741p x 0·161	441p
Rate per m	£4·41

Note: No formwork allowed for: concrete poured direct into trench.

Example 16

Concrete (1 : 3 : 6) 20 mm aggregate laid under 100 mm internal diameter drain pipe, the bed 440 mm wide and haunched up to top of pipe both sides.

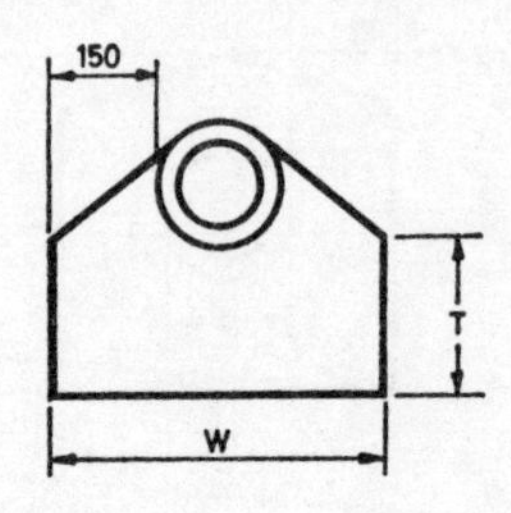

T = 100 mm for pipes under 150 mm diameter
150 mm for pipes 150 mm diameter and over

Consider 1 m length of pipe

Volume of concrete	0·44 x 0·10 x 1·00	0·044
	$\frac{1}{2}$/0·44 x 0·14 x 1·00	0·031
		0·075
Deduct	π/0·07 x 0·07 x 1·00	0·015
		0·060 m^3

Cost of 1 m^3 of (1 : 3 : 6) concrete 20 mm aggregate including profit and oncost as before — 2,741p

∴ 1 m costs 2,741p x 0·06 — 164$\frac{1}{2}$p

Rate per m — £1·65

20 Pro Rata or Analogous Rates

Occasions arise in agreeing Final Accounts where it is found that there are a number of items which do not form part of the original Bill of Quantities although they are similar and differ only in the quality of the material or slightly in the outputs. These items are normally dealt with on a *pro rata* basis.

Of the four main headings of (1) Materials, (2) Labour, (3) Establishment and Overhead Charges, and (4) Profit, it is possible within a unit rate to establish the cost of Materials and the allowance for Profit and Oncost, and thus isolate the Labour content. The Profit and Oncosts as stated by the contractor to have been allowed for within the rates can be checked by taking a number of *key* items from the Bill of Quantities and analysing to produce what should be a constant percentage. It is argued that the contractor would have difficulty in disguising this percentage if suitable items were selected.

Some people say that one can isolate the Profit and Oncost by estimating the Labour content. The writers believe that one might as well estimate the profit and oncost direct without going to the bother of using possibly questionable outputs.

BASIS OF CALCULATION

(1) Establish the percentage allowed for overheads and profit in the original rate and deduct the relevant fraction from the rate to reduce it to net cost.

(2) Isolate the unknown factors by calculating the part of the main rate which can be identified (e.g. material costs related to list prices).

(3) Adjust the unknown factors, if necessary by direct proportion.

(4) Add back the relevant allowances for materials (if applicable).

(5) Add profit and oncost.

Note: Care must be taken to ensure that rates can be calculated by *pro rata,* i.e. the work is carried out under similar conditions to that under which the original rate was priced.

INTRODUCTORY EXAMPLES

Example 1

Bill item

300 x 75 mm precast concrete cope weathered on top and twice throated on underside of projections m £3·85

Item in variation account

250 x 75 mm do. do.

300 x 75 mm = 22,500 mm² = 385p

$$\therefore 1\ \text{mm}^2 = \frac{385\text{p}}{22{,}500}$$

250 x 75 mm = 18,750 mm²

$$\therefore 1\ \text{m} = \frac{385\text{p}}{22{,}500} \times 18{,}750 = \quad £3{\cdot}21$$

This is not entirely satisfactory as the above assumes a constant labour output.

Example 2

Bill items

225 x 100 mm precast concrete slip sill, sunk weathered, throated, finished fair on three surfaces, girth 275 mm, reinforced with (2) 6 mm diameter mild steel rods cranked at ends m £5·16

275 x 100 mm do. do. m £6·20

Item in variation account

225 x 100 mm do. do. m –

225 x 100 mm =	22,500 mm²	
275 x 100 mm =	27,500 mm²	620p
Deduct	22,500 mm²	516p
Additional	5,000 costs	104p

$\therefore$ 1 mm² additional = $\frac{104\text{p}}{5{,}000}$

275 x 125 mm = 34,375
225 x 100 mm = 22,500

$\therefore$ Additional 11,875 costs

$\frac{104\text{p}}{5{,}000} \times 11{,}875$ 247p

Plus 516p

763p

Rate per m £7·63

Example 3

Bill items

(*a*) Excavates trench foundations not exceeding 2·00 m deep starting at ground level — m³ £3·85

(*b*) Excavates trench foundations not exceeding 4·00 m deep — m³ £6·54

(*c*) Basement excavation not exceeding 2·00 m deep starting at ground level — m³ £3·10

Item in variation account

Basement excavation not exceeding 4·00 m deep starting at ground level — m³ –

(*b*) £6·54 per m³
(*a*) £3·85 per m³

£2·69 for additional 1·50 m

It would be wrong to add 269p to basement rate of £3·10. The extra depth of basement excavation should increase in proportion to trench excavation.

Rate per m³ $\frac{310\text{p}}{385\text{p}} \times 654\text{p} =$ £5·27 (as against

310p + 269p = 579p)

Example 4

Bill items

(*a*)	Deposit surplus soil in spoil heaps not exceeding 100 m from excavations	m³ £1·10
(*b*)	Deposit surplus soil in spoil heaps not exceeding 150 m from excavations	m³ £1·36
(*c*)	Spread and level surplus soil not exceeding 100 m from excavations	m³ £1·68

Item in variation account

Spread and level surplus soil not exceeding 200 m from excavations — m³ –

(*b*) 136p
(*a*) 110p

26p for 50 m of additional wheeling.

(*c*)	
	168p
100 m @ 26 p per 50 m	52p
	220p
Rate per m³	£2·20

Example 5

Bill item

Excavate trench for fireclay pipe not exceeding 200 mm diameter, not exceeding 2·00 m deep and average 0·75 m deep including grading bottom, planking and strutting, filling in, compacting and disposing of surplus soil. — m £3·50

Item in Variation Account

Excavate trench for fireclay pipe not exceeding 200 mm diameter, not exceeding 2·00 m deep and average 1·25 m deep including do. — m –

Rate per m	350p
C/fd	350p

B/fd	350p	
Profit and Oncost 20% ∴ Deduct $\frac{1}{6}$	58p	
Net cost	292p	
Deduct Level and ram bottom $1{\cdot}00 \times 0{\cdot}55 = 0{\cdot}55\ m^2$ Labour 10 minutes per m^2 (agreed with contractor)		
∴ Cost $0{\cdot}55 \times \frac{10}{60} \times 190p$	17p	
	275p	
Excavating, planking and strutting etc. costs 275p for average 0·75 m deep		
∴ Excavating, planking and strutting for		
average 1·25 m deep = $275p \times \frac{1{\cdot}25}{0{\cdot}75}$		458p
Add back level and ram (for 1·25 m average depth)		
$0{\cdot}60 \times \frac{10}{60} \times 190p$		19p
		477p
Profit and Oncost 20%		95p
		572p
Rate per m		£5·72

Level and ram is 'taken out' of adjustment as it is a horizontal 'cost' whereas all other 'costs' are vertical and therefore more directly related to depth.

The following examples have been extracted from previous examination papers and are, of necessity, based on S.M.M. Fifth Edition (Metric) but relate closely to the Sixth Edition. However pro-rata rates on the Fifth Edition will be extant for some years. The rates have been up-dated to reflect 1977 rates.

EXAMINATION EXAMPLES

Example 6

Bill item

Concrete (1 : 3 : 6 mix) in beds 150 mm thick m^2		£4·00

Item in variation account

Concrete (1 : 2 : 4 mix) in beds 150 mm thick	m^2	£ –

Cement d/d site per tonne 2,400p plus unloading 152p = 2,552p
Sand d/d site per tonne 210p
40 mm whin aggregate 235p

(1 : 3 : 6) mix:	
1 cement @ 2,552p x 1·44	3,675p
3 sand @ 210p x 1·52	958p
6 aggregate @ 235p x 1·60	2,256p
	6,889p
Shrinkage Add 25%	1,722p
9 m^3	8,611p
∴ 1 m^3 costs	957p
(1 : 2 : 4) mix:	
1 cement @ 2,552p x 1·44	3,675p
2 sand @ 210p x 1·52	638p
4 aggregate @ 235p x 1·60	1,504p
	5,817p
Shrinkage Add 25%	1,454p
6 m^3	7,271p
∴ 1 m^3 costs	1,212p

Bill item

Rate per m²	400p
Profit and Oncost 20% ∴ *Deduct* $\frac{1}{6}$	67p
Net cost for 150 mm thick	333p
∴ 1 m³ costs $333 \times \frac{1{,}000}{150}$	2,220p
Deduct cost of 1 : 3 : 6 mix	957p
	1,263p
Add cost of 1 : 2 : 4 mix	1,212p
	2,475p
Profit and Oncost 20%	495p
Rate per m³	2,970p
∴ 1 m² at 150 mm thick = $\frac{2{,}970\text{p}}{1{,}000} \times 150$	446p
Rate per m²	£4·46

Example 7

Bill item

Concrete (1 : 3 : 6 mix) in foundations over 150 mm thick but not exceeding 300 mm thick. m³ £20·10

Variation account

Concrete (1 : 3 : 6 mix) in foundations over 300 mm thick m³ £ –

Bill item

Rate per m³	2,010p
Profit and Oncost *Deduct* $\frac{1}{6}$	335p
	1,675p

Mixer cost given as £3·15 per m³ Material from previous item 957p per m³	1,272p
C/fd	403p

Labour cost 403p

Taking output of labourer @ 190p per hour to be 2·12 hours per m^3
It will take less time to place concrete over 300 mm thick, say $1\frac{2}{3}$ hours (by negotiation with contractor)

$\therefore$ Labour cost = $\frac{1\cdot67}{2\cdot12}$ of 403p or $1\frac{2}{3}$ hours

@ 190p = 317p

Material costs	957p
Mixer costs	315p
Labour costs	317p
	1,589p
Profit and Oncost 20%	318p
	1,907p
$\therefore$ 1 m^3 over 300 mm thick costs	£19·07

Example 8

Bill item

One-brick wall in common bricks (1 : 3) cement mortar.	m^2	£14·25

Variation account

Brickwork in walls reduced to one brick thick in common bricks (1 : 3) cement mortar	m^2	£ –

Bricks £31·00 per 1,000 d/d site
Cement £24·00 per tonne d/d site plus 152p stacking
Sand 210p per tonne d/d site
Mixer £4·52 per hour output 1·15 m^3 per hour
Tradesman 220p per hour all-in rate
Labourer 190p per hour all-in rate

Mortar		
1 cement @ 255p x 1·44	3,675p	
3 sand @ 210p x 1·52	958p	
	4,633p	
Shrinkage add 25%	1,158p	
4 m³	5,791p	
∴ 1 m³	1,448p	
Mixer £4·52 ÷ 1·15	393p	
1 m³ costs	1,841p	
Bricks		
d/d site 3,100p per 1,000	3,100p	
5% waste	155p	
	3,255p	3,255p
Mortar		
0·60 m³ @ 1,841p		1,105p
		4,360p
1,000 bricks cost 4,360p		
∴ 118 bricks cost $\frac{4{,}360\text{p}}{1{,}000} \times 118$		515p
Bill item		
Rate per m²	1,425p	
Profit and Oncost 20%,		
∴ *deduct* $\frac{1}{6}$	237p	
Net Cost	1,188p	
Deduct Materials	515p	
Labour	673p	

Labour

118 bricks @ 673p @ 630p per hour of 2 tradesmen and 1 labourer gives output of 110½ bricks per hour.

Output should increase for reduced brick-work, say, 130 bricks per hour (by negotiation)

∴ Labour $\frac{630p}{130} \times 118$	572p
Material	515p
	1,087p
Profit and Oncost 20%	217p
	1,304p
Rate per m^2	£13·04

Example 9

Bill item
Extra over common brickwork for facing brick built Flemish bond and pointed as the work proceeds. m^2 £6·85

Variation account

(*a*) Extra over common brickwork for facing brick built English bond and pointed as the work proceeds.
Extra cost of facing bricks built English bond in place of Flemish bond is

$\frac{89}{79} \times 685p = £7{\cdot}72$ per m^2

(*b*) Extra cost of facing bricks built Scottish bond (3 courses stretchers to 1 course of headers) and pointed as the work proceeds.

Extra cost $\frac{74}{79} \times 685p = £6{\cdot}42$ per m^2

Example 10

Bill item

Extra over common brickwork for facing brick (P.C. £60·00 per 1,000 d/d site) in English bond, pointed as the work proceeds.	m^2	£7·72

Variation account

Extra over common brickwork for facing brick (P.C. £65·00 per 1,000 d/d site) in Flemish bond, pointed as the work proceeds,	m^2	£ –

Extra cost of new facings (£65·00–£60·00) × $\frac{79}{1{,}000}$	40p
Add for waste 5%	2p
	42p
Profit and Oncost 20%	8p
	50p
Rate for English bond is £7·72 per m^2 ∴ Rate for Flemish bond in same brick is $\frac{79}{89}$ × £7·72	685p
	735p
New rate per m^2	£7·35

Example 11

Bill item

Extra over common brickwork for facing bricks (P.C. £65·00 per 1,000 d/d site) built Scottish bond with (1 : 3) cement mortar and key-pointed with Snowcrete and shiver sand (1 : 3) at a later date.	m^2	£8·70

Variation account

Extra over common brickwork for facing
bricks (P.C. £58·00 per 1,000 d/d site)
built Scottish bond with (1 : 3) cement

mortar and key-pointed as the work proceeds.
On this particular contract it is established that the profit and oncost allowance is 12½%.
Pointing mortar costs £48·00 per m^3

Cost of facing bricks d/d site per 1,000	6,500p	
Stack: 2 hours @ 190p	380p	
	6,880p	
5% waste	344p	
	7,224p	
∴ Cost of 74 bricks = $\frac{7,224p \times 74}{1,000}$		535p
Cost of facing bricks d/d site		5,800p
Stack		380p
		6,180p
5% waste		309p
		6,489p
∴ Cost of 74 bricks = $\frac{6,489 \times 74}{1,000}$		480p

Rate per m^2		870p
Deduct Profit and Oncost 12½% = $\frac{1}{9}$		97p
		773p
Labour pointing 1 m^2 per hour (by negotiation) @ 220p =	220p	
Mortar 0·033 m^3 x $\frac{59}{1,000}$ x 4,800p	9p	229p
C/fd		544p

	B/fd	544p
Deduct Difference in cost of facings	535p	
	480p	55p
		489p
Add Key-pointing as the work proceeds 1·70 m² per hour @ 220p		129p
		618p
Profit and Oncost 12½%		77p
		695p

Rate per m² £6·95

Example 12

Bill item

One-brick wall built entirely of facing brick (P.C. £60·00 per 1,000 d/d site) built Flemish bond, both sides flush-pointed in cement mortar as the work proceeds m² £25·10

Variation account

One-brick wall built entirely of facing brick (P.C. £60·00 per 1,000 d/d site) built Flemish bond, one side flush-pointed in cement mortar as the work proceeds, other side struck-pointed in coloured mortar at a later date m² £ –

Coloured mortar £48·00 per m³

Rate per m²	2,510p
Deduct Profit and Oncost 20%; ∴ ⅙	418p
	2,092p
Deduct pointing one side: 1·70 m² per hour @ 220p	129p
C/fd	1,963p

		B/fd	1,963p
Add			
Labour 1 m^2 per hour 220p	220p		
Mortar 0·033 m^3 x $\frac{59}{1,000}$ x 4,800p	9p		229p
			2,192p
Profit and Oncost 20%			438p
			2,630p
Rate per m^2			£26·30

Example 13

Bill item

405 x 205 mm Welsh slates laid to 75 mm lap doubled-head nailed with 38 mm galvanized nails	m^2	£8·70

Variation Account

460 x 205 mm Welsh slates laid to 75 mm lap double-head nailed with 38 mm galvanized nails	m^2	£ –

Slates d/d site
405 x 205 mm £165·00 per 1,000
460 x 205 mm £215·00 per 1,000
Nails 55p per kg

Cost of 405 x 205 d/d site per 1,000	16,500p
Stack 2 hours @ 190p	380p
	16,880p
Nails 2 x 1,000 = 2 x 3·175 kg @ 55p	349p
	17,229p
2½% waste	431p
	17,660p

Covering capacity: $\dfrac{405 - (75 + 25)}{2} \times 205$

$= 0{\cdot}031{,}26\ m^2$ per slate
or $31{\cdot}26\ m^2$ per 1,000

$\therefore$ 1 m^2 costs $\dfrac{17{,}660p}{31{\cdot}26}$	565p	
Rate per m^2	870p	
Deduct Profit and Oncost 20%, $\therefore \frac{1}{6}$	145p	
	725p	
Material	565p	
Labour per m^2	160p	
Cost of 460 x 205 d/d site per 1,000		21,500p
Stack 2 hours		380p
		21,880p
Nails a/b		349p
		22,229p
$2\frac{1}{2}$% waste		556p
		22,785p

Covering capacity $= \dfrac{460 - (75 + 25)}{2} \times 205$

$= 0{\cdot}0369\ m^2$ per slate or $36{\cdot}90\ m^2$ per 1,000

$\therefore$ 1 m^2 costs $\dfrac{22{,}785p}{36{\cdot}90} = 618p$ say

If assumed that approximately same amount of time is expended per slate, labour costs would be dependent on area covered.

$\therefore$ Labour is taken as $\dfrac{31{\cdot}26}{36{\cdot}90} \times 160p =$ 136p

Material		618p
Labour		136p
	C/fd	754p

	B/fd	754p
Profit and Oncost 20%		151p
		905p
Rate is		£9·05

Example 14

Bill item			
21 x 70 mm white pine dressed facings rounded on two arrises		m	67p
Variation account			
(*a*) 21 x 95 mm white pine dressed facings rounded on two arrises		m	–
(*b*) 16 x 95 mm makoré dressed facings rounded on two arrises		m	–
(*a*) 21 x 95 mm white pine dressed facings cost £34·48 per 100 m			
(*b*) 16 x 95 mm makoré dressed facings cost £114·00 per 100 m			
Rate per m	67p		
Deduct Profit and Oncost 20% ∴ $\frac{1}{6}$	11p		
	56p		
Material			
21 x 70 mm facings cost £26·80 per 100 m			
∴ 1 m costs = 27p plus 5% waste	28p		
∴ Labour	28p		
(*a*) Labour taken to be the same			28p
Material			
21 x 95 mm white pine facing costs per m			
34½p plus 5% waste			36p
		C/fd	64p

	B/fd	64p
Profit and Oncost 20%		13p
Rate per m		77p

(*b*) Labour taken as 28p plus 50% say	42p
Material	
16 x 95 mm makoré facing costs per m = 114p plus 5% waste	120p
	162p
Profit and Oncost 20%	32p
Rate per m	£1·94

Example 15

Bill item

(*a*) 19 x 100 m red pine tantalized wallplates m 105 @ 45p
(*b*) 25 x 50 mm white pine dwangs between wallstraps m 390 @ 54p

Variation account

(*a*) 25 x 100 mm red pine tanalized wallplate m 90 @ –
(*b*) 38 x 75 white pine dwangs m 370 @ –

Red pine costs £164·21 per m^3
White pine 25 x 50 mm costs £14·45 per 100 m and 38 x 75 mm costs £32·59 per 100 m

(*a*) 19 x 100 m wallplate	45p
Deduct Profit and Oncost 20% ∴ $\frac{1}{6}$	7½p
Net Cost	37½p

Material
£164·21 per m^3
∴ 1 m costs £164·21 x 0·100 x 0·019 = 31p

plus 5% waste	32½p
Labour	5p

25 x 100 mm wallplate

Material
£164·21 x 0·100 x 0·025 = 41p plus 5% waste 43p
Labour 5p

48p
Profit and Oncost 20% 10p

Rate per m 58p

(*b*) 25 x 50 mm dwangs 54p
Deduct Profit and Oncost 20% ∴ $\frac{1}{6}$ 9p

Net Cost 45p

Material
25 x 50 mm costs £14·45 per 100 m, ∴ 1 m = 14½p plus 5% waste 15½p

Labour 29½p

38 x 75 *mm dwang*
Material £32·59 per 100 m, ∴ 1 m = 33½p plus 5% waste, say 34p
Labour for 25 x 50 mm dwangs is 29½p say output is $\frac{5}{6}$ of 25 x 50 mm dwangs
∴ $\frac{6}{5}$ x 29½p 35p

69p
Profit and Oncost 20% 14p

Rate per m 83p

Example 16

Bill item
Two coats Carlite plaster 13 mm thick on brick walls m² £2·08

Variation account
(*a*) Two coats Carlite plaster 10 mm thick on breeze partitions
(*b*) Two coats 13 mm thick on ingoes 100 to 200 mm wide

Carlite browning	3,550p per tonne	
Carlite finish		2,755p per tonne
Unload and stack		
$1\frac{1}{4}$ tonnes per hour		
@ 190p	152p	152p
	3,702p	2,907p

Material

1st coat: 11·1 mm thick

1 tonne browning covers 140 m^2

$\therefore 1\ m^2 = \frac{3,702}{140}$ $26\frac{1}{2}$

Finishing coat: 1·6 mm thick

1 tonne finish covers 410 m^2

$\therefore \frac{2,907}{410}$ 7p

$33\frac{1}{2}$p

Item rate per m^2	208p
Deduct Profit and Oncost 20%,	
$\therefore \frac{1}{6}$	$34\frac{1}{2}$p
(Proved from analysis of Bill rates)	
	$173\frac{1}{2}$p
Materials cost	$33\frac{1}{2}$p
$\therefore$ Labour costs	140p

(*a*) Material

Browning 8·4 mm thick,

$\therefore$ cost is $\frac{8{\cdot}4}{11{\cdot}1} \times 26\frac{1}{2}$p 20p

Finishing coat a/b 7p

27p

Labour

Same output agreed with contractor 140p

C/fd 167p

	B/fd	167p
Profit and Oncost 20%		33p
Rate per m^2		£2·00

(b) Material

Two coats	$33\frac{1}{2}$p
Labour	
140p plus 25% for narrow widths	175p
	$208\frac{1}{2}$p
Profit and Oncost 20%	$41\frac{1}{2}$p
	250p
$\therefore$ Rate per m = 250p x $\frac{200}{1,000}$	50p

Example 17

Bill item

30 mm cement and sand (1 : 3) screeded bed	m^2	£1·85

Variation account

40 mm cement and sand (1 : 3) towelled bed	m^2	–

Cement and sand 1,448p per m^3	
Rate per m^2	185p
Profit and Oncost 20%	
$\therefore$ *Deduct* $\frac{1}{6}$	31p
	154p
Material 1,448p x 0·030	$43\frac{1}{2}$p
Labour	$110\frac{1}{2}$p

40 mm towelled bed	
Material 1,448p x 0·040	58p
Labour output reduced by 50% allowing for difference between screeded and trowelled beds: $110\frac{1}{2} \times \frac{3}{2}$	166p
C/fd	224p

	B/fd	224p
Profit and Oncost 20%		45p
		269p
Rate per m²		£2·69p

Example 18

Bill item

16 mm wall plaster in 2 coats, the first coat of cement, lime, and sand (1 : 1 : 6) and the second coat of neat retarded hemihydrate plaster on brickwork finished with a trowelled surface. m² £1·90

Variation account

Do. do. 100 to 200 mm wide m –

Cement	2,400p per tonne	Sand	210p per tonne
Unload and stack	152p		–p
	2,552p		210p
Plaster	2,655p per tonne	Lime	2,800p per tonne
	152p	Unload and stack	152p
	2,807p		2,952p
Convert to lime putty 2 hours @ 190p			380p
			3,332p

1 cement 2,552p x 1·44	3,675p
1 lime 3,332p x 0·60	1,999p
6 sand 210p x 1·52	1,915p
	7,589p
Shrinkage add $\frac{1}{3}$	2,530p
∴ 8 m³ costs	10,119p
∴ 1 m³ costs	1,265p

Cost of m^2 is 1,265p x 0·0144 (as finish is 1·6 mm thick)	18p
1·6 mm hardwall plaster 2,807p per tonne covers 370 m^2 = $\frac{2,807\text{p}}{370}$	8p
	26p
Rate per m^2	190p
Profit and Oncost 20% ∴ $\frac{1}{6}$	32p
Net rate	158p
Material	26p
Labour	132p

New item do. do. 100 x 200 mm wide

Material	
Cement plaster	18p
Hardwall	8p
	26p
Labour	
132p plus 25%	165p
	191p
Profit and Oncost 20%	38p
	229p

∴ Rate per m 100 to 200 mm broad =

229p x $\frac{200}{1,000}$	46p

Example 19

Bill item

3 mm clear sheet glass in panes exceeding 0·10 but not exceeding 0·50 m^2 glazed to wood with putty	m^2	£8·75

Variation account

3 mm clear sheet glass in panes exceeding 0·50 but not exceeding 1·00 m^2 glazed to wood with putty | m^2 | –

Bill Item

Basic price of glass per m^2		472p
0·625 kg putty @ £10·20 per 50 kg		13p
		485p
5% waste		24p
		509p
Bill rate	875p	
Profit and Oncost 20%		
∴ *Deduct* $\frac{1}{6}$	146p	
Net cost	729p	
Materials	509p	
Labour	220p	

New item

Labour output increases by approximately 33% for larger panes;
∴ reduce labour cost by approximately 25%.

Labour		220p
Deduct 25%		55p
Labour		165p
Material price m^2		472p
Putty 0·50 kg @ £10·20 per 50 kg		10p
		482p
5% waste		24p
		506p
Labour		165p
	C/fd	671p

	B/fd	671p
Profit and Oncost 20%		134p
		805p
Rate per m^2		£8·05

Example 20

Bill item

(*a*) Prepare, one coat primer one coat undercoat and one coat gloss paint on plastered walls	m^2	£1·58
(*b*) Do. do. over 100 not exceeding 200 mm girth (This does not often occur)	m	45p

Variation account

(1) Prepare and two coats emulsion paint on new plastered walls	m^2	–
(2) Do. over 100 not exceeding 200 mm girth (This does not often occur)	m	–

Per 5 litres: Primer £4·50 Undercoat £6·50
Gloss £6·65 Emulsion £6·00

Materials
Covering capacity per 100 m^2
Primer: 10 litres @ £4·50 per 5 litres

$\frac{450p}{5} \times 10$	900p
Undercoat: 8·5 litres @ £6·50 per 5 litres $\frac{650p}{5} \times 8{\cdot}5$	1,105p
Gloss: 8 litres @ £6·65 per 5 litres $\frac{665p}{5} \times 8$	1,064p
100 m^2 costs	3,069p
∴ 1 m^2 costs	31p

Rate per m²	158p
Profit and Oncost 12½%	
Deduct $\frac{1}{9}$	18p
Net cost	140p
Material	31p
Labour costs	109p

At 220p per hour this means that it took $\frac{109}{220}$ hours = 0·50 hours say

A fair ratio of time between 2 coats emulsion and 3 coats oil is 4 : 9

∴ time spent on 2 coats emulsion would be 0·50 x $\frac{4}{9}$ = 0·22 hours

(1) *Material*

Emulsion 8 litres per 100 m² @ £6·00

per 5 litres = $\frac{600\text{p}}{5}$ x 8 960p

∴ 1 m² costs 9½p x 2 coats	19p
Labour 0·22 hours @ 220p	48p
	67p
Profit and Oncost 12½%	8p
Rate per m²	75p

(2) Rate per m	45p
Deduct Profit and Oncost 12½%$\frac{1}{9}$	5p
	40p per 200 mm broad
∴ Cost per m² = 40p x 5	200p
Deduct Material	31p
Labour	169p as against 109p (item (*a*))

∴ Approximately 50% increase in labour has been allowed by estimator.

Material	19p
Labour	
0·22 x 220p plus 50% additional labour	72p
	91p
Profit and Oncost 12½%	11p
Rate per m²	102p
∴ Cost per lin.m 100 to 200 mm broad	
$102 \times \frac{200}{1{,}000}$	20p

The above example, taken from one of the professional institutions' examinations, is considered by the writers to be outwith the limit of their own definition of *pro rata* but is included as a possible example for students' consideration.

21 Examination Questions

The following questions for the student's perusal and consideration are extracted from recent examinations set by The Royal Institution of Chartered Surveyors, the Institute of Quantity Surveyors, and the Scottish Association for National Certificates and Diplomas for H.N.D. courses in Building (which give exemptions from the Institute of Building examinations).

It is normal practice for material prices and on occasion the 'all-in' hourly rate to be given in the examination papers and this should be borne in mind when reading the following questions.

R.I.C.S.

(All questions would normally carry 20 marks)

1. The cost of employing labour is more than the basic hourly rate. Discuss the merits of the various methods of allowing in a tender for other costs and state what those costs are.

2. Define general overheads, project overheads and labour oncosts giving examples and stating how each is recovered by the contractor.

3. Build up the rate for the following item in a variation account:

   Two coats lightweight plaster 13 mm thick on brick walls exceeding 100 mm and not exceeding 200 mm wide } 50 m

   based on the following item in the bill of quantities:

   Two coats lightweight plaster 13 mm thick on brick walls } 180 m² £1·50

Labour output	Plasterer	Walls two coats	} 0·40 hour/m²
	Labourer		} 0·20 hour/m²

Price of materials (delivered to site)	Lightweight Plaster	– Browning (142 m^2/tonne) £28/tonne
		– Finish (460 m^2/tonne) £22/tonne

4. 'Risks and uncertainties facing the contractor, both in the estimates of the costs of the project and in the reactions of his competitors to the same tender situation, must play a large part in the assessment of the price quoted.' (Hillebrandt, P.M. Economic Theory and the Construction Industry).

   Discuss the above statement illustrating your answer with examples.

5. (*a*) State the factors to be considered when pricing the Brickwork and Blockwork section of a bill of quantities.

   (*b*) Build up the rate for the following item in a bill of quantities:

   Extra over common brickwork in 65 mm bricks for forming a flat arch in facing bricks 225 mm high on face 113 mm wide on soffit and pointed where exposed } 20 m

Labour output	Bricklayer	25 bricks/hour
Materials (delivered to site)	Common bricks	£18·00/1,000
	Facing bricks	£24·00/1,000
	Mortar	£13·00/m^3

6. Build up the rate for the following item in a bill of quantities:

   Remove excavated material to spoil heaps a distance not exceeding 220 m } 500 m^3

The contractor will use 1·5 m^3 diesel driven dumpers.
Cost £1,000 Interest $17\frac{1}{2}$ per cent.
Depreciation over 5 years. Site days 200 per annum.

Fuel consumption	2 litre/hour of operation
Diesel oil	10p/litre
Lubricating oil	25p/litre.

Loading is by a mechanical excavator at a rate of 10 m^3 'compacted soil' per hour and you are required to allow for the number of dumpers required to keep the excavator working at full capacity.

7. (*a*) Briefly describe the methods by which a tenderer may allow for his prices in the Preliminaries Section of a bill of quantities.
   (*b*) Build up rates for the following items in the Preliminaries Section of a bill of quantities assuming your own data for a contract of £250,000 and a contract period of 18 months.

   (i) Office for Foreman in charge;
   (ii) Safeguarding the works.

8. (*a*) Build up the rate for the following item in a bill of quantities.

   Excavate pipe trench for pipe not exceeding 200 mm diameter exceeding 1·50 m but not exceeding 3·00 m deep and average 2·50 m deep from ground level, including grading bottom, planking and strutting, filling in, compacting and disposal of surplus spoil } 15 m

Labour output	Labourer	
	Excavate	2·50 hour/m^3
	Extra throw	0·66 hour/m^3
	Fill, compact and spread and level surplus	1·50 hour/m^3
Materials (delivered to site)	Planking and strutting fixed	£0·75/m^2

   (*b*) Build up the rate for the following item in a bill of quantities:

   100 mm diameter vitrified clay pipes laid and jointed with tarred yarn and cement mortar (1 : 2) } 15 m

Labour output	Labourer	
	Laying and jointing	0·15 hour/pi

Materials
(delivered to site) 100 mm diameter
vitrified clay pipes £0·50/m
Tarred yarn £1·00/kg
Mortar mixed £13·00/m^3

9. The following item appears in the Joinery Section of a bill of quantities:

21 x 95 mm softwood skirting moulded one edge (no grounds) } 400 m £1·20/m

From this item prepare pro-rata rates for the following items in the variation account.

(*a*) 12 x 95 mm softwood skirting moulded one edge } 300 m
(*b*) 16 x 90 mm hardwood skirting moulded one edge screwed and pelleted (selected and kept clean for staining and polishing) } 100 m

Materials
(delivered to site) 21 x 95 mm softwood
skirting £0·65/m
12 x 95 mm softwood
skirting £0·40/m
16 x 90 mm hardwood
skirting £1·25/m

10. Build up rates for the following items found in a Bill of Quantities for a single storey reinforced concrete library block to a University value £80,000:

150 x 150 x 6 mm white standard quality glazed wall tiles to BS 1281 and bedding to plastered wall in tile adhesive with straight joints including grouting and pointing in approved grouting material
To walls ... ... 150 m^2
Fair square cutting against flush or projecting surrounds to openings 40 m
Raking cutting ... ... 5 m
Rounded edge ... ... 20 m

11. When employing labour on a building site the contractor has three choices of the method of their employment: (*a*) by employing labour direct; (*b*) by subcontracting on

a labour basis; and (*c*) by subcontracting on a labour and material basis.
Discuss fully on each event the factors which the contractor would have to take into account when pricing Bills of Quantities.

12. Discuss the factors influencing price levels in tendering.

13. Discuss and explain the adjustments that would be appropriate in negotiating the unit rates for the following items of concrete work based on a single-storey school building if a multi-storey addition is proposed for construction concurrently on site:

(i) Plain concrete (1:3:6) in foundation to trenches over 150 mm but not exceeding 300 mm thick ... ... m^3
(ii) Reinforced concrete (1:2:4) in beams over $0{\cdot}05\ m^2$ but not exceeding $0{\cdot}10\ m^2$ ... m^3
(iii) Reinforced concrete (1:2:4) in roof slab 150 mm thick ... ... m^2

14. Discuss the various cost factors that must be taken into consideration by a contractor when deciding whether to (*a*) buy, or (*b*) hire, the following items of plant.

(i) Backacter;
(ii) Central batching plant for concrete;
(iii) Dumper;
(iv) Tower crane.

15. The Bill of Quantities for a project contains approximately two-thirds of the contract value in new constructions and the balance in works of alterations and repairs. Discuss the differences that you might expect to find in the pricing of the alteration and repairs section of the Bill, as compared with similar items in the section of the Bill applicable to new works. Illustrate your answer with examples and give reasons for the differences.

16. The tendering policy of a contracting firm is decided by the Management. Discuss the factors that may influence the Management in submitting a tender figure for a contract which differs from the estimate compiled by the firm's estimating staff.

17. The following item appears in a priced Bill of Quantities:

    Tile roof with plain tiles, 226 x 165 mm,
    laid to a lap of 60 mm, double nailed, and
    including 38 x 18 mm softwood battens
    with 38 mm galvanized nails. ... ... m^2 £7·95

    From the foregoing, build up *pro rata* for the following items:

    Tile roof with interlocking tiles, 413 x
    330 mm laid to a lap of 76 mm, single
    nailed and including 38 x 18 mm soft-
    wood battens with 38 mm galvanized
    nails – per m^2

18. Illustrate the build up of an 'all-in' rate for:

    (*a*) a concrete mixer; and
    (*b*) define the cost of mechanical distribution of mixed concrete on a large housing site.

19. (*a*) Define project overheads and prepare an estimators list of such overheads to be considered for a local authority housing project.
    (*b*) Select *two* of the examples and calculate a hypothetical value for each with an explanation of your approach.

I.Q.S. (United Kingdom)

(All questions would normally carry 15 marks)

20. Build up prices for the following:

    (*a*) Vibrated reinforced concrete (1:2:4 – 19 mm aggregate) in casing to steel beam exceeding 0·10 m^2.
    (*b*) Ditto in casing to steel stanchion over 0·05 m^2 and not exceeding 0·10 m^2.
    (*c*) Ditto in bed over 300 mm.
    (*d*) Mild steel fabric reinforcement to BS 4483 Reference D49 weighing 0·77 kg/m^2 in wrapping to column.
    (*e*) Sawn formwork in 15 x 15 mm groove.

21. Build up prices for the following:

    (*a*) Excavate pit over 1·50 m and not exceeding 3·00 m

deep (assuming machine excavation) commencing at reduced level.

(*b*) Remove surplus soil from site.

(*c*) Softwood strut 38 x 38 mm between 150 mm deep joists.

(*d*) 150 x 50 mm softwood joists.

(*e*) 120 x 25 mm wrot softwood window board on and including softwood grounds plugged to concrete.

22. Calculate prices for the following alternative items:

(*a*) Ready-mixed mortar (1:1:6) and site-mixed mortar (1:1:6).

(*b*) One brick wall in commons in cement lime mortar (1:1:6) and 215 mm dense concrete block wall in cement lime mortar (1:1:6).

(*c*) 19 mm chipboard laid on battens (measured separately and 16 mm softwood tongued and grooved boarding in 100 mm widths fixed to battens (measured separately).

23. Build up prices for the following:

(*a*) 50 mm cement and sand (1:3) structural screed laid on precast concrete suspended floor, including workin around mesh fabric (measured separately).

(*b*) 50 mm ditto external paving laid on concrete with steel trowel finish in areas not exceeding 1 square metre.

(*c*) Knot, prime, stop and apply one undercoat and two finishing coats of oil paint on general surfaces of woodwork.

(*d*) Ditto surfaces not exceeding 100 mm girth.

24. Build up rates for the following:

(*a*) Formwork to sides of foundations (assume 10 uses). per m^2

(*b*) Ditto to horizontal soffit of floor (assume 6 uses). per m^2

(*c*) Ditto to 175 mm suspended edge. per m

25. Build up rates for the following:

(*a*) 125 mm P.V.C. gutter fixed to fascia with brackets at 1 metre centres and screwed to timber. per m

(*b*) Extra over last for 63 mm outlet. per No.
(*c*) 15 mm copper service pipe fixed with clips plugged and screwed to plastered brick walls. per m
(*d*) Made bend on ditto. per No.
(*e*) 350 mm x 250 mm white vitreous china lavatory basin complete with pair of 12 mm chromium plated pillar valves, 32 mm waste fitting, plug and chain and bed waste fitting in white lead and joint fitting to trap and joints to services, fixed with and including pair of white vitreous enamelled cantilevered brackets plugged and screwed to plastered brick walls. per No.

26. Build up rates for the following:

(*a*) 3 mm clear sheet glass to wood with putty in panes over 0·10 but not exceeding 0·50 square metres. per m^2
(*b*) 3 mm white figured glass group 2 ditto. per m^2
(*c*) 6 mm Georgian polished wired glass to wood with beads and glazing strip in panes over 0·50 but not exceeding 1·00 square metres including lining up in two directions. per m^2
(*d*) Glazing strip and embedding edge of glass in same. per m

27. The following rates have been extracted from a Bill of Quantities:

(*a*) Half brick skin of hollow wall in common bricks in cement mortar (1:3) in foundations. £3·25 per m^2
(*b*) 152 x 152 x 6 mm white glazed wall tiles bedded in cement and sand (1:3) on walls. £4·75 per m^2
(*c*) 13 x 75 mm Wrot softwood skirting plugged to plastered block walls. £0·75 per m

Analyse the foregoing and build up rates for the following:

(*a*) One brick wall in facings P.C. £45·00 per 1,000 No. delivered site in boundary wall in sulphate resisting cement mortar (1:4) finished fair face both sides and pointed as work proceeds. per m^2
(*b*) 152 x 152 x 6 mm white glazed wall tiles with one rounded edge on window sill 152 mm wide fixed with adhesive. per m

(*c*) 25 x 175 mm Wrot softwood door lining plugged and screwed to brick walls. per m

28. Explain the implications of the Working Rule Agreement in respect of a tender for building work.

29. Build up rates for the following:

(*a*) 25 x 250 mm pressure preservative treated softwood capping in one width to expansion joint plugged and screwed to concrete at 300 mm centres. per m
(*b*) 50 x 100 mm pressure preservative treated softwood splayed kerb. per m
(*c*) 35 mm (finished) flush door size 626 mm x 1,980 mm, covered both sides with 6 mm plywood and with hardwood edging strips tongued to two long sides. per No.
(*d*) Fix only cylinder night latch to softwood door and hardwood frame. per No.
(*e*) Fix only pull handle to softwood door. per No.

30. Build up rates for the following:

(*a*) 38 mm Granolithic paving (1 : 2½ mix) on concrete floors, including trowelling surface. per m²
(*b*) A fair flush edge 38 mm granolithic paving to other finishings. per m
(*c*) Granolithic skirting 75 mm high with small cone at bottom and with rounded top edge. per m
(*d*) Render and set in lightweight premix plaster on brick walls. per m²

31. Build up rates for the following:

(*a*) Prepare, knot, prime, stop and apply two undercoats and one gloss finishing coat on louvred wood door internally (measured flat overall). per m²
(*b*) Ditto on surfaces of woodwork over 100 mm but not exceeding 200 mm girth, internally. per m
(*c*) Prepare and apply by brush one mist coat and two full coats emulsion paint to plastered soffites, internally. per m²
(*d*) Prepare, seal and apply two coats of polyurethane lacquer on hardwood surface not exceeding 100 mm girth, externally. per m

H.N.D. (SCOTLAND)

(All questions would normally carry 20 marks)

32. (*a*) From current rates of wages build up 'All-in' hourly rates for tradesmen (brick layers) and labourers to include the charges which have to be covered by the estimator.
Allowance should be made for the following:

(i) distance to site 14 miles, daily fare 90p return
(ii) overtime worked for 2 days per week, 3 hours each evening
(iii) in each squad of 5 men one tradesman is paid 10p per hour extra as a working foreman

(*b*) Build up the unit rate for the following Bill Item: Hardcore bed 150 mm thick blinded on top with whin dust.

33. (*a*) The following item appears in the Concrete Section of a Bill of Quantities:

Concrete (1:3:6) in beds 150 mm thick m^2 £2·50

From this item prepare the pro-rata rate for the following item in the Variation Account:

Concrete (1:2:4) in beds 150 mm thick m^2

(*b*) List 3 factors which an Estimator must consider before pricing a Bill of Quantities explaining briefly the significance of each.

34. Build up the unit rate for the following Bill Items:

(*a*) 343 x 242 mm clay pantiles laid on and including battens to 56 mm head lap (from nail hole) and 38 mm side lap, each tile single nailed. m^2
(*b*) 19 mm roughcast in three coats having granite chip wet dash finishing coat (1:2) m^2
Note: 1st and 2nd coats 1:3
3rd coat 1:2

35. Build up the unit rate for the following Bill Items:

(*a*) 95 x 45 mm softwood door frame plugged to brickwork m

(*b*) 42 mm standard flush door type A No.
(*c*) 70 x 16 mm softwood door facing rounded on two arrises including mitres and ends m
(*d*) Supply and fix 100 mm butt hinges to softwood. No.

36. Build up the unit rate for the following Bill Item:
Excavate trenches for pipes not exceeding 200 mm diameter, exceeding 1·50 m but not exceeding 3·00 m and average 2·50 m deep; including grading bottom, planking and strutting, filling in, compacting and spreading and levelling surplus soil around trenches m
(Note: A 0·20 m^3 back acting excavator is to be used. Fuel consumption 2·5 litres per hour)

37. Build up the unit rates for the following Bill Items:

(*a*) BS Code No. 4 (19·52 kg/m^2) milled lead cover flashing 175 mm girth with 100 mm laps including turning into raggle and fixing with lead wedges (no allowance made in measurement for laps). m (10)
(*b*) Prepare and apply one coat primer; two coats undercoat and one coat gloss oil paint on general surfaces of woodwork. m^2 (10)

38 (*a*) Give FOUR examples of 'Overheads' and explain briefly how the allowance made to cover overheads in general is established. (4)
(*b*) The following item appears in the Slater Work section of a Bill of Quantities:

406 x 203 mm Welsh slates laid to 75 mm lap and double head nailed with 38 mm galvanized nails. m^2 £5·94

From this item prepare a pro-rata for the following item in the Variation Account:

355 x 203 mm second-hand slates laid to 75 mm lap and double head nailed with 38 mm galvanized nails. m^2 (16)

Index